Building the New Urbanism

The New Urbanism is perhaps the most influential movement that has emerged in suburban design, planning, and development in recent decades. It proposes to reform conventional suburban development by "building community." *Building the New Urbanism* asks "why new urbanism now?" to provide the first in-depth sociological investigation of the emergence of this phenomenon.

This volume situates the growth of the New Urbanism in the history of urban and suburban policy and development. The book builds an account of the movement's founding and development, linking its progress to the making of new places. The volume also investigates how the movement capitalized upon dynamics within architecture, planning, and the homebuying public to recruit support from among those groups. The book establishes a framework for analyzing the opportunities and constraints that confront any effort to change the way we produce the built environment. Moreover, it reveals how elaborately social the production of the built environment is and how specific the material solutions to social conditions must be to resolve this process.

Building the New Urbanism is an accessible volume that encapsulates and engages the dominant history of American suburbia. It draws on interviews with key figures, brings the work of prominent theorists of culture and science into the investigation, and broadens the focus of urban studies to the metropolitan region. It will be of particular interest to scholars and students of urban and suburban development, sociology, geography, and planning.

Aaron Passell is Assistant Professor of Sociology at Furman University. He specializes in the urban and suburban built environments, their development and redevelopment, and associated questions of social equity and environmental impact. His current research concerns urban historic districts and processes of neighborhood change.

Building the New Urbanism

Places, professions, and profits in the American metropolitan landscape

Aaron Passell

Routledge
Taylor & Francis Group
LONDON AND NEW YORK

First published 2013
by Routledge
2 Park Square, Milton Park, Abingdon, Oxon OX14 4RN

Simultaneously published in the USA and Canada
by Routledge
711 Third Avenue, New York, NY 10017

Routledge is an imprint of the Taylor & Francis Group, an informa business

British Library Cataloguing in Publication Data
A catalogue record for this book is available from the British Library

Library of Congress Cataloging in Publication Data
Passell, Aaron.
Building the new urbanism / Aaron Passell. -- 1st ed.
p. cm.
Includes bibliographical references and index.
1. Urbanization--United States. 2. Cities and towns--Growth.
3. City planning--United States. I. Title.
HT361.P377 2012 307.760973--dc23
2012013917

ISBN: 978-0-415-53897-8 (hbk)
ISBN: 978-0-203-10865-9 (ebk)

Typeset in Times New Roman
by Bookcraft Ltd, Stroud, Gloucestershire

This book is dedicated to all the people who have inspired me to undertake it and complete it, foremost among them, my mother, Anne, my son, Luca, and my partner in all things, Tabatha.

Contents

Figures

Figures

Acknowledgments

The intellectual substance of this project owes a great deal to Harvey Molotch, Neil Brenner, and Craig Calhoun, and Neil was also instrumental in helping get it published. In less direct roles, Rob Rosenthal, Charles Lemert, Vivek Chibber, David Garland, and Steven Lukes were all essential in my broader development as a sociologist.

Audra Wolfe provided invaluable assistance in shaping the book, as such. She was the latest in a number of thoughtful readers including Tabatha Abu El-Haj and Aaron Major. The form the book has taken was also influenced by anonymous reviewers for various journals and this press. In its last stages, Anne and Rob Goble provided careful criticism of the manuscript without which the book would not be what it is.

Much of the research was accomplished with support from the National Science Foundation's Graduate Research Fellowship, New York University's Henry Mitchell MacCracken Fellowship, and a Graduate School of Arts and Science Summer Grant. I also owe thanks to the other institutions and departments who sheltered me throughout a long process: the George Washington University Department of Sociology, Muhlenberg College Department of Sociology and Anthropology, and the University of Pennsylvania Urban Studies Program. Aspects of the project were also shaped by feedback at various stages at various conferences including the American Sociological Association, A Suburban World?, and NYLON. Robert Adelman, Alexandra Kowalski, Leonard Nevarez, and Melinda Milligan stand out in that crowd. I also never could have finished the research without the help of Roma Soriano.

This project would, of course, have been impossible without all of my cooperative informants. They are too many to name here, but Peter Katz and Andres Duany gave me a lot of time and important information, as did Jennifer Russel, Mark DePoe, and Richard Arkin. Payton Chang, at CNU, helped answer many questions. James Scott's *Seeing Like a State* (1998) originally set this project in motion.

I owe my ambitions for this project to role models who thoughtfully engage the world and argue about it, and who lead the way as professional academics, categories that are not mutually exclusive. I am lucky to count among these my grandfather, Lyne Few, my father, Nicholas Passell, and much of the family I married into – Thea, Nadia, and Samira, in particular. My parents, Anne and Rob, deserve mention here too, as does Tabatha, but so too do older and newer friends, Scott Rosenberg and Jeremy Teissere. Here's hoping that Luca finds a similar kind of inspiration, though for now it's hard to do better than Messi and Rondo.

1 Introduction

I grew up in a first-ring, streetcar suburb in an old Northeastern city. For me, the metropolitan built environment always seemed a dense, finely textured place. It integrated the histories, purposes, and buildings of the wide range of people who had once inhabited it and did then. It was a living landscape that combined incremental and catastrophic change with spaces that appeared to be almost static: store fronts became flats; historic buildings got modern facades, then lost them; some neighborhoods were demolished and replaced, while others were just left to age quietly.

The broader city, as I experienced it, was clearly a product of piecemeal negotiation, diverse interests, and wide-ranging, even conflicting, ideals. When I first encountered the New Urbanism, on the other hand, it seemed to be something else altogether. These new neighborhoods contained the density that one might expect in an urban space and familiar – old-fashioned – building types, but the underlying logic, or illogic, was missing. They were uniformly new, carefully organized – consistent – and clearly designed to convey some sort of meaning.

Certainly American cities contained instances of large-scale residential development – urban, subsidized housing projects being the most obvious example – but the New Urbanism's "traditional" design at scale mitigated the authoritarian impression those spaces gave off, by incorporating a kind of suburban banality. My initial readings on these spaces suggested that their builders maintained a conspicuous idealism that drew on a tradition of planned urban spaces in the United States and elsewhere. In the face of mixed historical evidence, the creators of these places believed they could shape the way people would live, that they could determine in advance, through design, the *meaning* of these spaces for their inhabitants.

As I began to investigate these new places, it quickly became clear that each was a product of its particular location and the moment in which it emerged. Yet all New Urbanist places shared features in common. Politics, economics, and culture at multiple scales all informed the designs and developments, but so did the individuals and professions involved in creating them. Less conspicuously social factors, like the standardized dimensions of the building materials, contributed to the look and feel of the developments. Then again, the introduction of actual inhabitants to these developments confronted – and, in some cases, swept aside – designers' intentions with the forces of everyday life. New Urbanist places were simultaneously and profoundly like and unlike other places in their origins and their relationship to the people who used them.

New Urbanism has only proved more popular in the decade since the movement first came to my attention during graduate school. New Urbanist developments are cropping up across the country and becoming increasingly common in the places where they have had the most success – the Southeast, Mid-Atlantic, Pacific, and Midwest states. The locations in which these developments seem to flourish are sprawling, de-centered regions, somewhere between suburbs and exurbs, lacking clear geographic focus – the Maryland and Virginia suburbs of Washington, D.C., and central Florida, for example. The New Urbanist effort to create "community by design" apparently strikes a particular chord in these areas and, while cultural critics object to an observed nostalgia, homebuyers have proven willing to pay a premium to escape conventional suburbia. As the total number of neighborhood-scale, New Urbanist developments across the United States approaches 1,000, we are forced to ask, "Why New Urbanism now?"

From a sociological perspective, the sheer hubris involved in "building community" is striking. New Urbanist designers are applying an understanding of the relationship between the built environment and social life that urban sociologists – along with urban historians, environmental psychologists, and others – still vigorously debate. If the movement were simply a forum for aesthetic or cultural reform, this certainty would be one thing, but it is more conspicuous when paired with concrete interventions in the suburban landscape.

The founders of the New Urbanism were galvanized to respond to the common form of American suburban development – incoherent metropolitan sprawl – by their observation of its inhumanity and increasing ubiquity. They see an American landscape dominated by the automobile, the result of both intended and unintended consequences of Euclidean zoning – the strict segregation of urban functions and their typical forms – and federal highway policy since World War Two. In response, the New Urbanists propose a comprehensive reintegration of the various functions of urban neighborhoods and pre-automobile small towns. While they acknowledge the impossibility of simply reproducing those traditional American small towns in the suburban now, they nevertheless attempt to reintroduce certain characteristics we have come to associate with urbanity and previous eras into the spatial relationships of suburban design, particularly walkability and a sense of community. They study the urban places they like best, that perform well according to their criteria, places like Savannah, Georgia; Charleston, South Carolina; Alexandria, Virginia; and Beacon Hill, Boston. Then they represent the relations they identify as responsible for these places' success in graphic, as opposed to verbal, "codes" at two scales – the urban or neighborhood and the architectural or design.

The intended effect of the combined urban and architectural codes is to produce a place that people want to use intensively and on foot. Some basic household needs – groceries, dry cleaning, video rental – should be accessible within a five- to ten-minute walk from any home in the neighborhood, as should schools, playgrounds, and ideally public transportation. Residents should enjoy these walks because of the appealingly proportioned "outdoor rooms" formed by the street, the sidewalk, and the exteriors of homes. Pedestrians should also appreciate the variety in design detail, the casual contact with neighbors encouraged by front porches and density, and the sense of "eyes on the street" (Jacobs 1961)[1] that adequate window area and the proximity of houses can suggest.

Yet this neo-traditional turn is selective in its attention to "walkability" and proximity: there is not yet a New Urbanist effort to do away with garages, just to move them, and no one in the movement seems to question the American middle-class homebuyer's apparent need for more than 2,000 square feet of living space. Moreover, the nostalgia embedded in the design of some of these places opens the New Urbanism to exploitation by non-New Urbanist developers who make superficial reference to their designs, capitalizing on nostalgia without incorporating New Urbanist spatial relations.

There are two primary New Urbanist project types: urban infill – intensifying and reconfiguring existing neighborhoods, ranging from HOPE VI redevelopment of federally subsidized housing projects to luxurious city homes; and suburban new towns – so-called greenfield development of an entirely new place, usually an expensive one. The HOPE VI program and its connection to the New Urbanist movement will be considered in Chapter 5, but this research is primarily focused on large-scale, new developments, because of their distinctness from their suburban surroundings and their consistency at significant scale.

By the end of 2004, at least 760 neighborhood-scale or larger projects were under construction or completed in at least forty-four states, housing as many as 12,000 people in a single project (*New Urban News* 2003; Steuteville 2005).[2] The growth rate since then has been such that the movement's primary press organ, the *New Urban News*, has been overwhelmed in its efforts to track projects and is developing a new tracking methodology. Economic analyses of New Urban new-towns have shown that homebuyers will pay as much as a 12–16 per cent premium over prices for comparable properties nearby, despite smaller lots (Song and Knaap 2003; Tu and Eppli 2001, 1999).

At the same time that it has become a fact on the ground, New Urbanism has asserted itself both as a collective and in the professions related to it – primarily architecture, planning, and development. The international Congress for the New Urbanism boasts "over 3,100 members in 20 countries and 49 states,"[3] and lists twelve state chapters, five "Chapter Organizing Committees," six "Interest Groups," student organizations,[4] and the affiliated Next Generation of New Urbanists. Major figures in the movement have assumed important positions at schools of architecture throughout the country,[5] while New Urbanist planning and design firms grow, proliferate, and win prestigious design prizes. The American Planning Association dedicated a "track" – one of six – entitled "The New Urbanism Comes of Age" to the movement's progress at its 2005 national conference in San Francisco – four full days of instructive workshops and reflexive panels. The Urban Land Institute, the professional organization for developers, publishes numerous how-to texts on New Urbanism as well as research on the movement and its products in their journal, *Urban Land*. "New Urbanism" means simultaneously and essentially an organized reform movement and the places that it has produced.

The New Urbanism in three takes

I have spent time in three developments that capture various aspects of the New Urbanism in particular ways. Seaside, Florida, is the first, proto-New Urbanist development – the movement would not get its name until several years after Seaside's completion. It established a key New Urbanist design firm, DPZ, and the key practices of the New Urbanist design process. Celebration, Florida, is a neo-traditional development outside of Orlando built by the Disney Corporation. It was neither designed by "card-carrying" New Urbanists, nor to New Urbanist standards, but it got so much publicity and is so frequently confused with early New Urbanist efforts that it deserves attention. The Kentlands, in Gaithersburg, Maryland, is, in some sense, the first fully realized New Urbanist, neighborhood-scale development and embodies many of the successes and shortcomings of the projects that followed it.

Seaside, Florida

As I arrived and struggled to park my large, American rental car – frustrated by signs for "Visitor Parking" that led nowhere and a great deal of curbside clearly marked "no parking" – I encountered the following sign:

> PEDESTRIANS ONLY
> No vehicles except
> OWNERS and their GUESTS

This sign was posted beside a street where Seaside proper abuts an older development called Seagrove. It captures a number of elements of Seaside: first, streets are for people, not cars. The exception is the "owners" of the predominantly second homes in Seaside coming and going from and to their first homes elsewhere. In my few days in Seaside, I saw many people walking between home and the beach, particularly at midday and the end of the afternoon, but golf carts were also ubiquitous. Seaside was clearly not a "town," as such, but rather a resort, structured in walkable relation to its focus and key public space, the sugar-white strip of sand along the Gulf of Mexico.

The second thing I noticed was the emphasis on ownership. Seaside is exclusive and expensive, beautiful and charming, but not welcoming to outsiders, at least not outside its commercial core. In this sense, it is a New Urbanist success, because its property values arise from its planning and design – as well as its location – which also seem to foster a territoriality that arises from sense of place. When I visited in 2002, housing prices started at $800,000 (Dunlop 2001). In post-recession 2010, zillow.com lists few properties with asking prices below $1 million and a number closer to $5 million (July 22, 2010). Seaside is wealthy and overwhelmingly white: just about any kind of population integration that we might associate with urbanity is impossible at those prices.

The development is tiny: approximately 350 lots, mostly single-family houses, on eighty acres –approximately twelve New York City blocks or eight in Chicago

– with a densely interconnected grid of streets, all of which lead either to the central square and commercial core or across the two-lane County Road 30-A to the beach. Porches and white picket fences are mandatory in the architectural code and the setbacks from the street – where houses are placed on the lot – are minimal. The resulting effect is that you always feel close to those inside when you walk, even though the boundary between public and private space is clearly marked. Narrow sand paths, for foot traffic only, cut through the middle of blocks.

The architectural code for the development specifies a great deal about the houses and lots: the appropriate landscaping and foliage; the pitch, symmetricality, and material of the roof; the kind of wood to be used in siding; that "Individual fence patterns shall not replicate another on the same street" (Mohney and Easterling 1991: 261). The code mandates the proportion and styles of doors, the proportions and styles of windows, and driveway surfaces. Nevertheless, the code does not specify exactly what the houses should look like, and each seems slightly different from the next.

But even if Seaside is a second-home, resort community that excludes many of the functions and people we identify with urbanity, it is a lovely place. Begun in the early 1980s, it is the first concrete product of proto-New Urbanist design principles. It is located well west on the Florida Panhandle, a stretch of coast once referred to as the "Redneck Riviera" but now marketed as Florida's "Emerald Coast." It is closer to Alabama, Georgia, and Mississippi than most of the population centers in Florida, and, anecdotally – counting license plates of cars parked by houses on a July afternoon – draws on those states for most of its homeowners and visitors.

It was designed by Andres Duany and Elizabeth Plater-Zyberk, two of the founders of the Congress for the New Urbanism, for developer Robert Davis, after a now legendary tour of the Southeast that identified the key elements of the local vernacular and the underlying spatial relations of Southern, small-town urbanism. The next step was a charrette – an intensive, interdisciplinary, iterative, problem-solving, design session over multiple days that incorporated as many pertinent professionals and local stakeholders as possible. The Seaside charrette produced both the urban code, determining the layout of the development and the arrangement of various uses in relation to one another, and the architectural code, mentioned above, specifying building design in the interest of its performance in shaping and inflecting space.

Duany and Plater-Zyberk were early in their careers at the time, having recently left Miami firm Arquitectonica to found DPZ in 1980. Both had received undergraduate degrees in architecture from Princeton, then proceeded to Yale's architecture program for Masters degrees, where they had worked with, among many others, Vincent Scully, prominent architectural historian and scholar of American urbanism. The social network they developed over the course of their education would constitute much of the foundation of the future movement. The couple's design for Seaside would launch the collective effort to derive and analyze desirable urban qualities and build them into new places, in response to the concurrent depredations of suburban sprawl. Davis, the developer, became a partner in the Arcadia Land Company and has since built a number of New Urbanist projects.

Celebration, Florida

While my immediate and instinctive response to Seaside combined envy with outrage at conspicuous privilege, my initial reaction to Celebration was something closer to fear of the dark. On a muggy Wednesday evening in late July, this early exemplar of neo-traditional planning and design felt more like a neo-ghost town. In an hour-and-a-half of walking neighborhood streets, I encountered only three people: a woman walking a dog, clearly made nervous by my presence; a female jogger who overtook me from behind so noiselessly as to make me jump; and a teenage boy glimpsed as he crossed a street-lit intersection far ahead, no longer visible by the time I reached it. The streetlights, widely spaced and alternating between sides of the street, left portions of the sidewalks dark, punctuated by a brightness disconcerting in its contrast to the rest of the street. Hundreds of little anoles, the ever-present, central Florida lizard, scampered around my feet as they moved from the street-side grassy strip to the front lawns on the other side of the sidewalk.

On paper, Celebration is organized around a vibrant commercial core. From my visit, however, the design of the neighborhoods raises more interesting issues. The walkability – and the community nominally generated by it – associated with neo-traditional design is often moot in the central Florida climate. When I visited in July, it was still hot at 9:00pm and the air felt like you could swim through it. Most reasonable people were indoors and even the peeps of countless tree frogs couldn't drown out the powerful hum of the ubiquitous cooling systems. All of the houses were closed up tight and no cheerful evening sounds – conversation or even television – made their way as far as the sidewalk. Needless to say, by mid-afternoon the next day, the atmosphere was stifling.

In contrast to Seaside, where each house was individually commissioned in accord with the architectural code, in Celebration houses must be built in one of six styles: Coastal, Classical, Colonial Revival, French Normandy, Mediterranean, or Victorian – which combination of styles has never occurred organically in any one place. Each style is carefully described in the Celebration Pattern Book, a neo-traditional reinvention of an historical builders' reference. Only two builders were responsible for all of the project's early residences (Frantz and Collins 1999: 80–1). This combination generates a kind of uniform, historical cuteness executed in modern materials.

But like Seaside, house styles must vary within a block – no house can be the same style as the one on either side of it. House colors are selected from a limited palette associated with their respective style, and may not be the same color as any other within three houses in either direction, except if the color is white (ibid.: 65–6). The result is a Disney-ish reproduction of a pre-World War Two, small-town neighborhood that never existed, at quite a large scale – approximately 9,000 residents in eight distinct neighborhoods as of 2008, projected to reach 20,000 at build out.[6] There is little of the reference to local vernacular architecture that characterizes Seaside; no elements of the design were intended to deal with the difficulties of this particular climate. Instead, there is a modernist obstinacy that technology can make houses designed for other climates work here.

Accordingly, in addition to ubiquitous air conditioning, I passed vinyl siding creeping with mildew and stucco cracking in the combined heat and humidity.

Celebration was developed by the Disney Corporation in the early 1990s – the first residents arrived in 1996 – near Orlando, on land that Disney already owned and over which it exercised extensive political control. It stands out for the scale of its ambition and the resources committed to its realization – Celebration avoided the financial problems that plagued other early New Urbanist projects. It also attracted a great deal of attention: two books on the community were published within a few years of its opening, one by a prominent American Studies scholar, the other by established journalists (Ross 2000 and Frantz and Collins 1999). The development's association with Disney has created some complications for living there, in that visitors to the amusement parks sometimes also visit Celebration's shops and restaurants and occasionally fail to recognize the public/private boundary – a number of houses near the town center display signs that would be unnecessary elsewhere reading "Private Home, Do Not Enter," "Private Residence," or "Home Occupied, Please Stay Off Property."

Celebration bears a complex relationship to the emergence of the New Urbanism. Despite being one of the first, and certainly one of the most prominent, neo-traditional developments to achieve occupancy, none of the movement's leaders were centrally involved. It was planned and designed by Cooper, Robertson and Partners and Robert A.M. Stern. Both Jacquelin Robertson and Stern have evinced a sympathy to the New Urbanism and have won New Urbanist design awards, but neither has been active in the movement. However, the Celebration Pattern Book – the style guide for all housing –was created by Urban Design Associates, whose co-founder and principal emeritus, Raymond Gindroz, was an important New Urbanist early adopter.

The differences between Celebration and the New Urbanism run deeper than personnel. Several aspects of Celebration's design and planning run counter to New Urbanist principles. A sign of its neo-traditionalism, rather than its New Urbanism, is that Celebration's nostalgia is conspicuous and intentional, concerned with explicit historical reference. The specification of styles, particularly generic ones from elsewhere in the country, rather than local vernacular, violates the New Urbanist emphasis on creativity within guidelines. The importance of the distinction here – neo-traditional vs. New Urbanist – is that the New Urbanism has struggled to get the difference recognized: the conflation of the two has exaggerated accusations of nostalgia against the New Urbanists and the association with Disney has proved problematic in elite cultural circles.

The Kentlands

By contrast to Seaside and Celebration, the Kentlands nearly feels like an authentic place – a privileged, homogeneous place in the D.C. suburbs, but a place nevertheless. I witnessed groups of actual small children walking home from school on a sunny spring afternoon. There is a diner to walk to in the neighborhood commercial center, ice cream and coffee, and an ice skating rink in winter. Moving out into the neighborhoods, there are parks and playgrounds, a recreation center complete with swimming pool, and two small "lakes" with walking paths around them – the man-made lakes are part of the storm water management system of the development. Unlike Seaside and Celebration, the Kentlands is a straightforward archetype of New Urbanist, greenfield, neighborhood-scale development.

The Kentlands was designed in 1988 through a collaborative process led by Andres Duany, founding New Urbanist and co-principle of DPZ, for developer Joseph Alfandre. Construction was begun in 1990 and, by 2001, the 352-acre development was substantially complete, including approximately 1,800 homes, numerous parks, an elementary school, a mixed-use, "main street" shopping district with restaurants, offices and a multiplex cinema, as well as a large, conventional shopping center at the edge. By 2007, the 340-acre, adjacent Lakelands – once part of the same property – was also substantially complete. Lakelands is designed according to the same principles as Kentlands, by the same firm, but not to as demanding a standard of materials and traditional design intervention.

The Kentlands' designers toured the mid-Atlantic region to develop a sense of the local vernacular style before envisioning the development. They chose to reproduce it quite faithfully, rather than depart from it: brick and clapboard dominate, with pitched roofs of cedar shingles and sash windows with divided lights – in modern materials, of course. Homes are mostly two and three stories. The density drops off from the center: the shop awnings of the main street have apartments above, the cross streets have attached town houses, then smaller single-family homes slowly give way to larger ones as you move further into the residential neighborhoods. Homes are close to the sidewalk, enclosing the space of the street. Streets follow the contours of the somewhat hilly terrain, provide ample sidewalks shaded by trees and parallel parking, and frequently intersect with other streets. Public buildings and parks punctuate the various neighborhoods, so that pedestrians often have a clear visual goal.

While the Kentlands succeeds in some respects – it looks and feels like many of the older, small towns nearby – it fails in others. The original proponents of New Urbanism aspired to a degree of income and racial integration that is simply not possible in a predominantly upper-middle class Maryland county (Montgomery) that serves as a bedroom community for Washington, D.C.. On one of my first visits in 2004, a house had just sold for $1 million. Anecdotally, everyone I encountered in the Kentlands was white. Moreover, the walkability of the development and the integrated commercial functions do not serve as employment centers for most of its residents. While the dentist and the realtor may be able to afford to live in Kentlands, the majority of the businesses are low-wage services, like nail salons, dry cleaners, and restaurants. Most people who live in the development must drive elsewhere to do their white-collar jobs. Some hold out hope for public transit, but it has yet to be developed.

Locating the New Urbanism in the history of planned development

In the sense that it tries to create a daily-life-encompassing, built environment that is envisioned whole, the New Urbanism is not unique. In its attempt to address social conditions by building, the New Urbanism inserts itself into a tradition of planned development whose utopianism, uncompromising totality of approach, and reliance on the state have colored related efforts over the past century. The New Urbanism partially defines itself in contradistinction to these characteristics, taking instead a pragmatic, incremental, and pro-market stance.

American architect Frank Lloyd Wright and Swiss architect Le Corbusier are archetypal for the design and planning professions, primarily for the scope and scale of their visions and their claims to personal authority. Working early in the twentieth century, they both espoused a physical determinism that led them to address a broad range of the world's ills through wiping the existing landscape clean and beginning again, though each with different emphases. Wright sought to encourage American democracy through the deconcentration of habitation, while Le Corbusier thought to rationalize the city by segregating its functions. Their legacies can be found in sprawling American suburbs and the land-use zoning that shapes modern American metropolises. At the same time, their personal authoritarianism relates to the aspect of planning of which Americans are most suspicious.

Wright and Le Corbusier were idealistic and demanding in their attempts to create wholesale change and their most profound interventions – new cities – were never realized as such. Less ambitious, smaller scale efforts to improve suburbs arose in the United States in the 1920s and 1930s with more lasting effects. Radburn, New Jersey, for example, introduced the cul-de-sac to suburban planning. Other efforts, like the Greenbelt Towns program of the New Deal, established the possibility of building new places in the United States, but highlighted the political limitations of government involvement in real estate development in a market economy. The for-profit American new towns of the 1960s and 1970s – Reston, VA; Columbia, MD; The Woodlands, TX; Irvine, CA – did not confront the same obstacles, but were also each individually one-offs, backed by major financial interests.

The complications of the relationship between designers, planners, government institutions, and markets in the United States have had important consequences for the professions and the landscape. For example, planners lost a great deal of autonomy and authority after federally subsidized, urban renewal-related housing projects failed spectacularly, in part because they were implemented through complex processes that compromised essential aspects of the original planners' intentions in the interest of the market and local control (Hirsch 1998). Historical efforts to plan the built environment have accreted not only in the physical landscape, but also in the professions that undertake design and planning, defining their cultures and their ways of knowing. That means that the New Urbanism has confronted the legacy of these efforts both in the logistical sense of what has worked and what has not, but also as embedded in professional practice, a legacy that sometimes facilitates innovation and change and at other times frustrates them. Historical efforts at planned development therefore constitute both opportunities – particularly in the sense of examples to draw upon – and constraints for their contemporary successors.

Defining terms

The detailed arrangement of the physical spaces we inhabit shapes our lives and informs them, both by literally constraining our actions and by communicating layers of cultural detail both to and about us as we move through the world. Our goal for understanding the built environment should be essentially structurationist (Giddens 1984): to show how social structure and individual action are mutually constituting, or, in our case, to envision the built environment and social life in a dynamic and reproductive relationship. We need to focus on the potential for change that lies within ongoing processes of development, on the practices of the actual players in the process of producing the built environment, and on the specific materiality of the places produced – how it captures and reproduces the commitments of all those involved.

In describing a landscape of opportunity and constraint, I am drawing on well-established uses of those concepts. McAdam's (1999) use of "opportunity" implies not only that opportunities emerge as a result of broad historical processes, but also that they can be actively expanded or reconfigured by the actors to whom those opportunities present themselves. McAdam also asserts that a degree of organizational capacity and consciousness of group interests needs to exist for actors to successfully exploit emerging opportunities – which do not activate themselves or alter the landscape independently of concerted action by individuals. Molotch (2003) describes the constraints upon the development of new material objects similarly. The dynamics of an historical instant define the opening for a new thing to come into being, but those dynamics can be influenced by the people trying to create that thing. Molotch's work makes clear, furthermore, how critical the specific materiality of the solution is to how successfully it overcomes constraints.

The specificity of the form that the urban built environment takes contains the resolution of tension among broad structures and individual actions. That specificity can produce consequences both intended and otherwise. Any example of built environment can only be the way it is because of the convergence of its planners, architects, developers, builders, funders, marketers, and inhabitants, their sense of how to accomplish that which they pursue and their particular efforts, and their confrontation of historical trends and forms. Elements of the physical make-up of the city and their relation to one another represent unique and uniquely material solutions to complex social problems of interests and interactions, professions and politics, traditions and expectations. The myriad, subtle details of a design can grab and hold all of the abstract elements together, can make them "interactively stabilized."

The production of the built environment is occurring simultaneously at at least two significant levels of abstraction. At the concrete level, places are built by laborers, carpenters, developers, to the designs of architects and planners, within the permitted uses of local governments, according to financial analyses of funders, to be purchased by consumers. Places embody direct, material processes of production. But, as should be obvious, all of the concrete processes occur within a context defined at a higher level of abstraction, primarily by professional fields with varying intellectual, aesthetic, political, and economic commitments and each with internal dynamics of its own. To do justice to the study of the built environment as social process, then, we need to account for both concrete and abstract practices.

Vision and field

Distinguishing the concrete from the abstract aspects of the built environment and its production is essential. It is equally important to understand the process in terms of conception and execution: the built environment must be conceived or envisioned before it can be executed or constructed.[7] Both conception and execution are socially located activities, characterized by their embeddedness in systems of knowledge, aesthetics, and practice.

More than distinguishing conception and execution as stages of the process of the built environment's production, we can understand them as in tension. The built environment as a social process is the resolution of a tension between "vision" and "field." These are the most important characteristics of conception and execution, respectively. By vision, I mean the imaginary place that exists first in one mind, but ultimately, necessarily if not identically, in the minds of all who participate in the project in any capacity. By field, I mean an abstract social space,[8] actively bounded and characterized by a comprehensive set of dispositions that substantially shape existence within it, from forms of knowledge to appropriate practices, both constraining and motivating action. Both of these concepts will be elaborated below.

Tension denotes the conflict between what we can undertake guided by our mind's eye and the demands that real, if abstract, social structures place upon that undertaking. Resolution, here, means both realization and compromise – a particular built environment originates in the individual imagination and proceeds through an intricate process in which the image interacts with the requirements of the real world, rarely emerging in concrete form precisely as abstractly conceived.

The generative claim of this research, then, is that one social process that constitutes the built environment is the realization of a place through compromise demanded by the confrontation of the creative imagination with the dynamic constraints of the many-layered abstract social space within which it originates and pursues concreteness.

Real estate development – the production of the built environment – entails the transformation of a narrowly held vision – like a discovery – into a broadly recognized built environment through the involvement of a diverse group who necessarily share the vision, in variations related to their interests, in their common struggle for realization. Latour (1987: 108) refers to this sort of transformation as "translation" – the process of making something meaningful to others – and argues that it involves both "enrolling" a group to carry the vision forward into facticity and controlling that same group, so that they produce a fact as faithful to the original vision as possible.

Like the technological projects of Latour's concern, a real estate development project remains a fiction until it is fully lashed-up, inhabited (Latour 1996: 23).

That fictitious nature, the imaginary quality of the vision, the imprecision of its communication, allows the malleability that translation and enrollment require. "Only finished projects have an essence" (ibid.: 48). Some version of the original vision needs to become the fact the team is reproducing, the solution to their respective problems of planning, profit, and a place to live.

Latour writes, " … the easiest means to enroll people in the construction of facts is to let oneself be enrolled by them!" (1987: 110). In other words, rather than confronting groups of potential allies with the discovery of an outsider and trying to persuade them by demonstrating the essential merit of the discovery, the instigator should work to define a new group that includes himself and those he would like to recruit, a group whose goal is "translated" into the construction of the fact in question, because its facticity serves their interests (ibid.: 115, 117). If the appropriate reinterpretation of their interests is provided, allies will encourage the originator to join their efforts, bringing his new information with him, and thereby reifying the discovery.

A field is a bounded social space within which action is organized, both constrained and encouraged, by common affective intuitions – appropriate goals are recognized, as are the methods acceptable for their attainment. A profession, like architecture, can define a field, but the "field" should not be confused with the merely disciplinary actions of training or qualification. Architects do not participate in the field of architecture simply by getting a degree and then getting paid for design work. They do architecture by acting "like architects" and thinking "like architects," knowing the sorts of things that architects know – and, presumably, ignoring the sorts of things that architects don't. They do not do architecture any other way. The field does not dictate or delimit behavior so much as motivate and guide practice, nudging it in the right direction and setting up a common context within which to exercise individual ingenuity.

We internalize the demands of the fields we inhabit so that they shape not only what we believe is appropriate to do, but what we can imagine is possible and how we do that imagining. And by being this way, we reproduce the field itself. Doing architecture like architects communicates what architecture is and how it is properly done. Similarly, but different, planners reproduce planning and bankers banking by doing them just so, which is very much not like doing architecture.

The reproduction of the field guarantees us some stability and continuity, sometimes privilege, and usually recognition. The field buttresses our ability to claim expertise and resist incursions from outside it. Doing architecture the way that architects do it reproduces both the organization of the social space of architecture and its boundedness, maintaining an area of expertise and the distinction between that area and what is outside of it. Nevertheless, the field is not static: shifts in practice will alter the field as it is reproduced, both changing goals and methods and the area over which expertise is claimed.

The development of field-specific *habitus* – the internalization of the dispositions inherent in the field – in interaction with an actor's position within the field, promotes particular practices. We no longer have to be concerned with how practices reflect interests and how to sort out personal and professional influences on interest. Instead, we can see that actors both perceive field-specific avenues for action and are simultaneously impelled by that perception.[9] Practice is not conscious calculation, of potential gain or anything else, but action that begins "below the level of explicit representation and verbal expression" (Bourdieu 1985: 728).

The belief or sense associated with a field is practical in that it is embodied, incorporated, pre-conscious. It is enacted and re-enacted, much like Latour's built fact, incorporated in practice (Latour 1987). Habitus is not like a rulebook frequently consulted, but rather like the fundamental presupposition of a sport – in soccer, we don't use our hands. Violations may be clarified, codified, and consciously referred to, but the essential condition holds – we play with our feet, legs, chests, and heads, however we intuit doing so, and when we don't we are not playing soccer.

Outline of the book

Part of understanding the now in "Why New Urbanism now?" requires distilling the pertinent historical conditions from an existing history of post-World War Two urban and suburban development in the United States. This will be the subject of Chapter 2. The postwar suburban real estate boom gave rise, in turn, to a set of conditions critical for the story of the New Urbanism's emergence: among them, changes in planning education, the failure of urban renewal, and the absence of architects from a developer-driven design process. These conditions, among others, constituted the opportunities upon which the New Urbanists would capitalize in bringing their movement together.

The movement itself began as an informal network of architects with a shared educational background. Chapter 3 begins by reconstructing the New Urbanism's founding through the words of its founders. It goes on to explore the Congress for the New Urbanism and its charter, but also argues that the organizational progress of the movement was interdependent with the development of New Urbanist places. The Kentlands, in Gaithersburg, Maryland, is a key early project and makes an excellent case.

Chapter 4 is focused on how the management of opportunities and constraints characterized the New Urbanism's emergence. The New Urbanists recruited planners and architects by developing a new, empowering, interdisciplinary design process – the charrette – and by turning zoning codes upside-down, while preserving them – in form-based codes. They addressed environmentalist criticisms of conventional suburban development and the perceived absence of community by emphasizing compactness and walkability in their designs. Finally, the movement attempted all this while simultaneously struggling to meet the demands of the private market for housing and the requirements of real estate finance.

Having seen the New Urbanism emerge, we need to consider how it survived. Chapter 5 asks: What sort of ideological response did this reform effort generate and how did it manage resistance from within the design and planning professions? In addition, we consider how the New Urbanism reinforced its position through prominent interventions: its involvement with HOPE VI, the reconstruction program for federally-subsidized public housing; and, its response to the destruction wrought by Hurricanes Katrina and Rita.

Chapter 6 explores the inhabitation of New Urbanist places and the contribution of homebuyers to the constitution of the movement. Returning to the Kentlands case study, we examine how the project's champions negotiated various obstacles to its completion. The sense of community and the pursuit of it emerge as key factors in the selection into and the inhabitation of New Urbanist places, but what community involves proves complicated. Despite its best intentions, the New Urbanism seems to facilitate the self-segregation of the privileged in the metropolitan region.

The Conclusion returns to the social-theoretical claim that motivates the research. This multi-dimensional portrait of the New Urbanism has demonstrated the manner in which the production of the built environment is an elaborately social process. Not only do we need to expand our understanding of the physical space of cities and suburbs beyond that of a static container for social life, but we also need to incorporate dynamic conceptions of the interactions among politics, economics, culture, and professions to explain how it is that the material and social affect one another. This book establishes a framework for doing so that should be applicable to other cases.

2 Before the New Urbanism

Introduction

The New Urbanism embodies a qualitative shift in the suburban built environment. In moving away from automobile-dependent, residential subdivisions toward walkable, mixed-use neighborhoods based on an older model, it is simultaneously a shift in the form of the places being built and in underlying social and professional structures. New Urban projects get realized when they successfully *redirect* the efforts of those involved in the construction and consumption of suburbia in its conventional form. To understand this shift, we have to establish the predominant conditions *away from* which this shift has occurred. That is the task of this chapter. While the broader New Urbanist project includes urban infill in addition to suburban greenfield development, the perpetuation of the conventional suburban form was the original impetus for New Urbanist reform.

The New Urbanism is dialectically related to post-World War Two suburban development: the conditions that gave rise to what has become conventional suburban development, gave rise to the conditions out of which the New Urbanism constituted itself. These conditions defined the terms in which the New Urbanism would have to express itself, provided it with the impetus it needed to launch itself, and constrained it to operating in a particular fashion to sustain itself.

More specifically, federal mortgage, tax, and transportation policy, firm composition in the homebuilding industry, elite and popular portrayals of urban and suburban cultures, and professional formation and practice in planning and design combined to define the context out of which the New Urbanism would emerge. To understand this, we must consider how each phenomenon was produced and reproduced and how it came to pose a problem or inspire complaint in a way that elicited attempts to solve or resolve it. The broader context of postwar metropolitan development can, in other words, be seen retrospectively as a complex of problems to which the New Urbanism offered a response.

Much of what I discuss below will be familiar to those who know the leading histories of suburban development in the post-World War Two period (Fishman 1987; Hall 1988; Hayden 2003; Jackson 1985; Kruse and Sugrue 2006). My application of these analyses, though, will emphasize a resistance to suburban growth beginning in the 1970s, rather than American suburbs' origins in the 1930s and 1940s – or earlier still. I will draw on these histories to describe a specific social and material space in which the New Urbanism could establish itself. I will demonstrate below both how the conjunction of factors that shaped the suburban landscape opened opportunities for the constitution of the New Urbanism and constrained the movement to responding comprehensively to those factors. Further, I will establish how the conditions that developed within the pertinent professional fields in the same period similarly constrained and enabled the emergence of the New Urbanism.

This chapter begins with a discussion of the immediate postwar period – the housing shortage the nation confronted coming out of the war, the policies implemented to remedy it, and their consequences, intended and otherwise. Next, we consider the nearly-contiguous moment in the 1950s, distinctive for reinforcing the explosion of suburban homebuilding and buying by constructing highways, segregating land uses, and consolidating a kind of suburban sub-culture. After that, we will examine what I call the post-postwar period of the 1970s and 1980s, exploring how the conditions of the postwar period evolved into those out of which the New Urbanism would arise. These evolving tendencies include: bigger houses on more land, a secondary market in mortgages, more mall construction and traffic, and environmentalism as an outgrowth of suburban culture. We shift to an in-depth discussion of the professions integral to the suburban development process, planning and architecture, and the dynamics within them. I discuss, in particular, changes in planning education and practice, the absence of architects in conventional suburban development, the failure of urban renewal, and its consequences for the planning profession. The chapter concludes with a review of how the conditions enumerated here have operated as opportunities and constraints that have importantly influenced the New Urbanism's emergence.

The post-World War Two suburban boom

Establishing conventions, sowing the seeds of crisis

The factors driving post-World War Two suburban development in the United States established a specifically American form of suburbanization that was refined and expanded into the 1970s and 1980s, though it was neither qualitatively altered nor significantly challenged up to that time. In fact, it proceeds unabated today in many parts of the country. Seven significant, interrelated factors – federal mortgage policy, the rise of big builders, interstate highway funding and construction, federal tax policy, a divergence of attitudes toward suburban culture, urban renewal, and suburban municipal exclusionary zoning – contribute to the particular form of postwar suburbia. Each of these interrelated factors is critical to establishing the conventions of suburban development, but the combination also contains within it the causes of its own transcendence.

From shortage to boom to sprawl

Post-war suburban development in the United States was so concerted and so intense that it created a qualitatively new environment, American suburbia, and a new way of producing it, large-scale community builders. This was, in major part, because of the scale of the demand for housing to which it was responding.

Homebuilding was already slow throughout the Great Depression before a moratorium on non-war related construction between 1941 and 1945 eliminated private homebuilding completely (Hall 1988: 294). Extensive military housing and war-worker housing was constructed in strategic areas throughout World War Two, but the real estate lobby insisted that the majority of it be destroyed, rather than permanently benefiting its residents (Hayden 2003: 130), to make certain the federal government stayed out of the business of direct housing provision. At the end of the war, according to Hall, between 3.25 and five million families were inadequately housed (1988: 294). This shortage was quickly exacerbated by the baby boom that followed the return of American servicemen. The federal government was confronting a housing shortage of unprecedented proportions.

Hall's figure implicitly raises the question of how we define "adequate" or appropriate family housing, but that question has long had a single answer in American culture: the detached single-family home, owned by its inhabitants, is the ideal (see particularly Fishman 1987). The federal government cooperated with real estate and development interests to implement that cultural and political-economic ideal. They did so by revolutionizing mortgage finance in a way that left large-scale developers in the best position to capitalize upon it.

Buying the suburbs

The Home Owners Loan Corporation (HOLC) (Jackson 1985: 195), legislatively created in 1933, and the Federal Housing Administration (FHA) (ibid.: 203), part of the National Housing Act of 1934, became important after World War Two. They established both the structures of modern mortgage finance and federal mortgage insurance, the provision of housing through the private market, and the resulting anti-urban, racialized, suburbanizing tendencies. Neither these financial structures nor their consequences should be regarded as the only possible result of the conditions they confronted, rather they should be seen as solutions that satisfied the demands of the players in the process with the loudest voices.

The HOLC and the FHA rationalized and standardized a diverse collection of lending practices for real estate by establishing and popularizing long-term, self-amortizing mortgages. In effect, this rationalized and standardized both the production and consumption of housing and radically broadened the market. Modern mortgages represented a large step toward affordability from the previously typical packages – five to ten year loans that were not entirely repaid when the final settlement was due, necessitating renewal, risking foreclosure, and carrying a significant stigma through the nineteenth century (Jackson 1985: 196). The FHA, and later the Veterans Administration (VA), insured mortgages – guaranteeing their repayment to the lender – thereby eliminating much of the risk to for-profit lenders of providing mortgages to working or lower-middle class homebuyers, and thereby making it more profitable. This reduction of risk also made mortgages cheaper for the borrower – meaning more people could buy homes for less and with less.

Together, these programs consecrated the American commitment to the provision of shelter through the private market. This was a conscious commitment (see Hayden 2003) and something of a departure from the norm among advanced economies. According to Jackson, publicly owned housing represented 46 per cent of the market in England and Wales and 37 per cent in France in 1980 (as compared to 1 per cent in the United States) (Jackson 1985: 224). Anton explains that in the 1960s, 65 per cent of housing in Stockholm was publicly built, two-thirds of the rental housing was out of the private market, and the Swedish state supplied 90 per cent of housing loans (Anton 1975: 138, in Logan and Molotch 1987: 147–48). Moreover, 65 per cent of new housing in Great Britain was provided by public agencies between 1945 and 1960 – declining to 45 per cent in later decades (Pahl 1969; Rex and Moore 1967, in Logan and Molotch 1987: 148).

This commitment to private homeownership in the United States has had profound consequences for the process of production of the built environment of the metropolitan region and its specific form. Foremost among them is that housing is produced for profit – and is designed, by the current era, to hold its value. No significant alternative to established development tendencies can even get its footing without first demonstrating the potential for a substantial return on investment. In addition, alternatives to conventional development must contend with the homogenizing force of the homebuying finance mechanism.

Big builders

As well as making borrowing cheaper for homebuyers, federal mortgage insurance programs allowed developers to borrow money against the promise of federally insured mortgages for the housing they were constructing. Big builders – firms capable of building tens or hundreds of houses simultaneously – were best able to take advantage of this opportunity, because they could achieve economies of scale vis-à-vis both paperwork and production. By the late 1950s, such large-scale operations were producing "about two thirds of the new houses in the United States" (Hayden 2003: 132).

Levitt and Sons was the best known of these "community builders" (Weiss 1987), but builders in every major American city took up Levitt's assembly-line-like production techniques (Jackson 1985: 238). They constructed thousands of houses at a time by building virtually the same house throughout developments in a carefully routinized process that broke the work down into relatively simple steps. This process was deskilling, thereby saving labor costs, but also capitalized on efficiencies in design – house dimensions reflected standard material sizes – and the vertical integration of the supply chain. These techniques dictated the uniformity and repetitiveness of new, suburban, residential architecture in the 1940s and 1950s (ibid.: 239).

Large postwar developments were planned and built, not as neighborhoods, but as aggregations of a single house, a fact that prospective buyers could only come to experience in time as residents. This phenomenon is both exemplified by and explains important qualities of Levittown's marketing and layout. Levitt and Sons marketing materials are notable for only ever showing one house at a time, blurring off into vagueness somewhere near the lot lines. Moreover, the views of these lone houses are from distances and at angles that could only have been accomplished through the exercise of significant artistic license on the part of the renderer, given the dimensions and density of the real places (Liell 1952). This was encouraged, according to Hayden, by the cultural idealization of the single family home, referred to above. She writes, "Sometime early in the twentieth century ... Americans began to idealize the house and separate it from its neighborhood" (2003: 190).

Valuing homes, valuing homebuyers

Federal mortgage insurance programs and the popularization of mortgages required the development of property-appraisal methodology that could establish the value of the house being mortgaged and insured. This methodology emphasized comparability and quantifiability. This emphasis is aptly described by James Scott as "seeing like a state" (Scott 1998), by which he means reducing diverse elements of the social landscape to bureaucratically manageable elements. Accordingly, real estate appraisers assessed not just the houses and properties themselves, preferring new, single-family homes, with minimum lot sizes, setbacks, separations, and widths (Jackson 1985: 206, 208), but also their surroundings. Appraisals took "neighborhood stability" into account when assessing the investment's value as collateral. Neighborhood stability was virtually synonymous with racial homogeneity and thus made the integration of new neighborhoods and inter-racial proximity in established areas virtually impossible (Jackson 1985: 207). Further, the very presence of African American residents was presumed to indicate declining property value, thereby establishing the practice of "redlining" and institutionalizing this self-fulfilling prejudice.

The suburban boom democratized homeownership in the United States for some, because the combination of inexpensive housing and housing finance made homeownership available to people in the working and middle-classes who could not previously have afforded to buy a house. It was a critical element of the process of becoming White for many Jews and European immigrants (see, for example, Brodkin 1998), enabling them to blend into a mass consumer movement (Cohen 2003). At the same time, it was just another step in the disenfranchisement of Black Americans. Redlining affected lending, but, more than that, many developers – Levitt and Sons among them – explicitly refused to sell to Blacks. A number of scholars assert that the exclusion of Blacks from this mid-century opportunity to build equity is key to the radical Black-White wealth disparities that persist today (Conley 1999).

Similarly, neither FHA nor VA mortgage insurance articulated an explicit anti-urban bias, but they obviously favored a kind of residential development that was primarily practical in the suburbs. This meant that housing finance flowed to the suburbs and not to cities. The relative disinvestment in cities was obscured by the magnitude of the suburban boom and the optimism of the immediate postwar era, but would ultimately produce effects that demanded attention (see my discussion of urban renewal in the section on professionals below).

Thus did a housing shortage and a baby boom get transformed into a suburban boom through the intricacies of federal mortgage policy, big builders' construction practices, and American cultural ideals of home. According to Hall, the 1960 Census showed that, "while the central cities grew by 6 million or 11.6 per cent [between 1950 and 1960], the suburbs grew by a dizzy 19 million, or by 45.9 per cent … the decade of the greatest suburban growth in American history" (Hall 1988: 294).

Reinforcing the suburban boom

The segregation of land uses and the rise of the mall

In its ultimate influence on the form of postwar suburbia, the rationalization of land use was as important as the regularization of housing. After World War Two, local zoning regulation rationalized land use by dis-integrating and clustering traditionally integrated uses, residential and commercial in particular, thereby promoting the legibility of the landscape and enshrining the conditions and demands of a particular moment – especially for automobility – with little thought for the mutability of those conditions.

Zoning had been in use since long before the postwar suburban boom: it was first conceived to separate noxious uses from residential areas in industrial metropolises. That aspect of land-use zoning rapidly became less pertinent as the pace of American de-industrialization increased through the 1960s and 1970s (Bluestone and Harrison 1982). Nevertheless, the strict geographic separation of land uses both locally and at the scale of the metropolitan region became more thorough in the postwar period, extending even to the segregation of multi-family from single-family residential building types.

As with the system of real estate appraisal, use segregation contributed to the stabilization of property values by simplifying developed areas. Zoning reinforced the preferences of real estate finance by mandating conforming building types, lot layouts, and housing-to-lot-size ratios – properties became increasingly likely to resemble those around them. This mandated conformity was frequently put to nefarious purposes: "exclusionary zoning" requires properties to perform according to standards that price them out of reach of lower and middle-income households, thereby excluding them from places so zoned.

Since new residential development in the postwar period was so strictly residential, separate spaces needed to be built in which the millions of new suburban residents could feed, clothe, and entertain themselves. The federal government encouraged the development of now ubiquitous strip shopping centers and regional malls by changing tax law in 1954 to allow the accelerated depreciation of commercial property (Hayden 2003: 168). Previously, new commercial construction had been considered to lose its value for tax purposes over the course of forty years of use. The legislated acceleration of that process compressed forty years into *seven*, meaning that the proportion of a new building's cost that could be used to offset income associated with ownership of the property suddenly got much larger, at least in the short term. This meant that much less of that income was subject to tax, much more of it profit. Over 98 per cent of malls built thereafter were profitable for their investors (ibid.).

Accelerated depreciation had a profound effect on the location and form of shopping centers and malls by making them short-term investments, best executed on the least regulated territory at the lowest design, construction, and materials costs. They were built in peripheral locations, further into the metropolitan hinterlands than existing residential development, where they faced the absolute minimum of planning constraints, in terms of roadway capacity and construction standards. Relatedly, their useful life as physical entities rapidly came to conform to their useful life as taxable entities: commercial spaces were not built to last much longer than the period of their depreciation. Thus, the 1950s, 1960s, and 1970s produced a profitable overabundance of low-quality commercial space.

Tax law thus reinforced the land-use regulatory tendency toward the disintegration of uses by fostering the continuous new construction of extensive commercial spaces. The widespread construction of malls also attacked existing integrated spaces – downtown shopping districts – and rendered many of them unsustainable. Moreover, while zoning concentrated commercial functions geographically, accelerated depreciation – and related commercial real estate tax law (see Hanchett 1996) – concentrated the construction of commercial space professionally, helping to create the mall developer, as such.

The new road-building regime

The HOLC, the FHA, and VA mortgage insurance unleashed a torrent of postwar homebuilding and buying in outlying areas. Accelerated depreciation encouraged an explosion of mall construction. Compounding these effects, the Interstate Highway Act (IHA) of 1956 flooded the metropolitan region with a new kind of traffic. Urban public transportation systems, streetcars in particular, suffered, some by neglect and some by design, as automobile transport gained in popularity. Furthermore, road building often functioned as a tool used to eliminate nuisance neighborhoods and nuisance populations – by which was meant the poor, particularly immigrants and people of color.

By 1943, the American Road Builders Association, a broad-based pressure group made up of oil, rubber, asphalt, and construction interests, automobile manufacturers and dealers, truck and bus companies, banks, advertising agencies, and labor unions – "a lobbying enterprise second only to that of the munitions industry" (Jackson 1985: 248) – began the effort to build bigger, better, faster roads and to have the government pay for them. A more secretive group called the "Road Gang" maneuvered the 1944 Federal Highway Act into place, getting 60 per cent federal funding for local road building (Hayden 2003: 165). The 1956 IHA raised that percentage to ninety and mandated the collection of gasoline taxes, dedicating the revenues exclusively to highway construction (Jackson 1985: 250) that by 1958 had cost more than $41 billion (Hayden 2003: 166).

All of this money dedicated to roads built interrelated networks at various scales. While the interstate system is clearly essential to the industrial and commercial history of the nation, intra-urban and intra-metropolitan road networks are more relevant for the areas of our concern here. Modern, local road networks are *dendritic* – like the branches of a tree or the veins of a leaf – smaller thoroughfares leading into larger ones, then larger ones still. This frequently means that there is only a single route from one suburban location to another, but also means that all drivers trying to get from neighborhoods to shopping centers or office parks and back traverse the same local collectors. Finally, this network presumes and reinforces automobility – walking is no longer an option due to distance and lack of sidewalks.

Sub(urban)-culture?

The explosion of new suburban homes in the postwar period and their uniformity and consumption by an eager mass encouraged a strong reaction in the intelligentsia to a presumed emptiness of suburban culture. Academic works like William H. Whyte's *The Organization Man* (1956) and David Riesman's *The Lonely Crowd* (1961) connected the physical form of suburbia and its proliferation to a decline in community and changes in "social character." Literary Johns Cheever and Updike portrayed a bleakness hidden beneath the suburban dream.

This negative response from the cultural elite defined half of a strictly bifurcated response. While the elite looked down upon suburbia and suburban life, the broad middle of American society leapt at the chance to own their own homes in the new developments. Herbert Gans's *The Levittowners* (1967) disputed studies like Whyte's and Riesman's by arguing that life in these new places did fulfill many of its promises, suggesting there was no simple truth of the matter. Moreover, film and television content with popular appeal, from "Leave it to Beaver" to "The Brady Bunch," reproduced suburban life in its best light at least until the 1980s.[1]

Evolving tendencies

More house, less household

Neither the radical postwar housing shortage nor the baby boom were enduring phenomena. Unprecedented rates of residential construction and rapid suburbanization in the 1940s and 1950s successfully addressed the need for more family housing and made the United States a suburban nation by 1960. This meant a commitment to a particular metropolitan form in the sunk costs of infrastructure, which were significant, but also a more widely shared culture of housing and home, a new set of expectations for a greater segment of the population. The continued availability of relatively affordable land in metropolitan areas, the extent of which is unmatched elsewhere in the world, has meant that a mass-production development model could be reproduced for decades with minimal friction. In other words, the rapid suburbanization of the 1940s and 1950s established trends that would not simply fade away as the urgency of the crisis to which they responded declined.

The indefinite geographic extension of the metropolitan area has enabled an emphasis on new-ness in housing and a continuous – if punctuated – increase in consumer expectations, both originating in the immediate postwar period. On the one hand, five- to ten-year-old houses are considered harder sales in the suburban market. On the other, once single-family houses were established as an attainable ideal, lawns got bigger, houses got bigger, then they incorporated all manner of enhancements, from central air conditioning to open plans to, by now, granite countertops in stainless steel kitchens. New suburban houses and their typical characteristics set the standard for the housing market.

By the late 1970s, according to Jackson, high interest rates and rapid inflation were simultaneously pushing new home prices out of reach of many young families and pushing new home construction further out in search of cheap land and profit – consuming three million acres of farmland per year (Jackson 1985: 284). This rapid conversion of land was hastened by two tendencies identified by Hayden: the increasing size of the average single-family house and lot; and the decreasing size of the household – the destabilization of the Ozzie and Harriet household model, which itself had replaced a multigenerational family. In 1950, near the height of the postwar suburban boom, the average new house was 800 square feet on a 5,000 square foot lot, with two bedrooms and one bath – and some unfinished attic space – intended for a married, heterosexual couple and their three children. By 1999, the average new house was 2,250 square feet on a 12,910 square foot lot (Hayden 2003: 190), but median house size peaked in 2006 at 2,268 square feet and had declined to 2,100 by 2009 (Emrath 2010: 2). Household size had decreased, meanwhile, and household composition, at least among the homebuying population, had scattered among families with fewer children, single parent families, childless couples, and non-family groups of co-habiting adults. Hayden reasons,

> The smaller the household, the more houses were needed to contain the population. The bigger the houses and yards, the longer the distances became. The amount of land used to house Americans expanded much faster than population growth would have suggested.
>
> Hayden 2003: 190

This increase in average house and lot size involves increases in housing cost and the demand for developable land, but more importantly reflects changes in expectations of space per household member – it sets new standards. Nineteenth century, American pastoralism (Fishman 1987; Hayden 2003) has become, over the intervening century, a yearning for the MacMansion. This is more important than the specifics of house or lot sizes: the suburban ideal has come to be embodied foremost in individual homes, rather than in a broader, more inclusive sense of the built environment – the neighborhood, for instance. These expectations have proven durable and shift only in particular circumstances under concerted pressure.

Requiring repetition: the secondary market in mortgages

The federally administered mortgage programs that succeeded FHA and VA mortgage insurance and grew out of them demanded an even greater uniformity of housing product, with the impetus for uniformity shifting from production and appraisal processes to financing. The Federal National Mortgage Association (Fannie Mae) and the Government National Mortgage Association (Ginnie Mae) created a standardized mortgage instrument that allowed banks around the country to channel savings into fast-developing regions. This led, in turn, to a secondary market in mortgages and their securitization: mortgages were bundled – as assets that provide a revenue stream – and then shares in the bundle were sold as publicly tradable stocks.

For mortgages to be sold, bundled, and traded, they have to be consistent and comparable products. This means they rely on homes of similar value for collateral and are issued to borrowers of similar likelihood to repay the loans – in theory.[2] Both of these post-FHA phenomena thus further exaggerated the unvarying repetition of proven design and planning formulae in the suburbs and their racial and socio-economic homogeneity – though the explicit inclusion of race in property value assessment was eventually made illegal.

Of course, from a post-Great Recession perspective, the securitization of mortgage debt ultimately exaggerated the democratization of homeownership irresponsibly, actually taking advantage of some of the people who had been left out of the original suburban boom through predatory lending, and made a mockery of financial institutions claims about their requirements. More on the effects of the Great Recession will follow in later chapters.

Malls, later

By the late 1970s, approximately 22,000 suburban shopping centers had been built (Hayden 2003: 170). The cost to the federal government of accelerated depreciation was between $750 million and $850 million per year in the late 1960s and over a billion dollars per year in the 1970s (Hayden 2003: 164, citing Hanchett 1996). The Tax Reform Act of 1986 made mall construction less profitable, but left suburbs in a curious position. More than enough commercial space existed to serve the needs of suburban residents into the 1990s, but much of that space would deteriorate before then and be decommissioned, a large-scale case of planned obsolescence. Changes in mall design and accompanying customer expectations – atria, food courts, indoor amusement parks – in new malls farther out into the distant suburbs and the arrival on the scene of the big-box category killer – such as Target – hastened this process.

Stuck in traffic

The dendritic hierarchy of thoroughfares that defines the modern metropolitan road network aggregates traffic in such a way that any increase in potential users – either through new development or an increase in car ownership per household (in which the road system and metropolitan spatial distribution are themselves complicit) – demands an increase in capacity at the middle levels of the system, because any additional throughput is necessarily channeled through existing routes. The conventional response to demands for increased capacity of major arterials is increased road width. Increases in road width carry the immediate and obvious financial costs of construction and the less easily measured, but profound experiential costs to local users, both because of the inevitable lag between demand and provision and because of the frustrations experienced during construction. In fact, a phenomenon observed in traffic studies and referred to as "capacity-induced demand" suggests that even increased road capacity rarely increases travel speeds in congested areas. In other words, the decision to rely primarily on cars to move people in and around the suburbs has lead to more cars and greater need for them, even as it has proven difficult to manage that increase.

All of this is to say that, by the 1980s, Americans had gotten sick of sitting in traffic. Before everyone needed to drive everywhere, driving everywhere seemed like a good idea. Jane Holtz Kay estimates that, by 1997, Americans spent eight billion person-hours per year stuck in traffic (Kay 1998: 14), while the Urban Mobility report argues that hours lost to traffic peaked in 2006 at 5.25 billion, declining to 4.62 billion by 2009 (*The Week* 2011). It is the combination of auto dependence and the lack of alternatives – both in terms of routing, because of the form of the road system, and in terms of mode of transport – that has driven the conventional suburban form to its traffic crisis.

Few convictions are as deeply entrenched in American culture as the "right to drive" and the ongoing romance with the car. Many Americans simply will not reconsider automobile dependence, but that dependence is also a vicious cycle. The infrastructure of automobility is so well developed and the suburban landscape so structured by it, that any alternative seems inefficient and inflexible and is quite expensive. Seemingly stuck in their frustration with traffic, many American suburbanites have become willing to consider development alternatives.

Suburban culture shift: sprawl and environmentalism

Hayden asserts that what might have been a general anti-suburban sentiment among the 1950s and 1960s intelligentsia got reconstituted as an "anti-sprawl" critique in the 1970s and on (Hayden 2003: 194). This critique ranged from attacks on fringe development and consumerism in the West by the photographers of the "New Topographics" exhibition (1975) to a federal government published report by the Real Estate Research Corporation, *The Costs of Sprawl* (1974), revisited and republished in 1998.

This discursive shift made a new alliance possible. While the elite critique was directed at a lifestyle the masses thought themselves fortunate to attain, the two groups could hardly speak. But when the elite critique was reframed to focus on the aspects of the conventional suburban phenomenon most conspicuously problematic – traffic in particular – criticisms with which the residents could identify and could hardly disagree, it encouraged cross-class cooperation. By the 1980s, this cooperation had made sprawl an important campaign issue that transcended party lines (Hayden 2003: 195).

While it was ultimately "spun to the right in the late 1990s" (ibid.), the early fight against sprawl was led by environmental organizations like the Sierra Club and the Natural Resources Defense Council (Hayden 2003: 194). General impact fees were briefly levied against developers, in the hopes of forcing them to pay for some of the "externalities" of new development – infrastructure costs and new constraints on access to previously public goods (Hall 1988: 420). The legality of some such fees was overturned by the US Supreme Court in 1987, because their relation to documented impact was uncertain (ibid.).[3]

Another anti-sprawl reform from this period has endured, though: the Environmental Impact Report (EIR). Increasing concern for the environment through the 1960s led to the National Environmental Policy Act of 1969 (NEPA). NEPA required, among other things, that executive-branch departments of the federal government obtain EIRs for federally funded development projects. The EIR is intended to specify the relevant benefits and costs of a given project (Logan and Molotch 1987: 162). The federal requirement was adopted and expanded as a procedural device by many states in the 1970s, covering a wide range of public and private development, and reveals the popularity of a rising environmentalist critique of development in that period.

Though it has evolved considerably since the 1980s, the EIR frequently has a number of limitations: the typical analysis is narrow in scope, with the loss and benefits considered accruing primarily to local government – as opposed to residents in general or some more abstract, systemic "environment"; costs are calculated for the project at hand, not the cumulative effects of all related development; liabilities are hard to quantify; mitigation is problematic; and, there is always the possibility of invoking overriding considerations – jobs, housing, tax base – (Logan and Molotch 1987: 163–4). At the same time, the exposure and analysis of such issues can affect the process, as was originally intended. EIRs tend to encourage development and, once ruled upon, can allow almost anything, but they do afford local publics procedural rights in the planning review process (Logan and Molotch 1987: 162–4).

Contemporary attitudes toward suburban culture and development have made space for resistance to conventional development in regulatory politics. At the same time, as we will see below, the original elite/mass divide endures, undermining possible reforms. First, things "suburban" remain culturally subordinate, for the most part, below the consideration of high-minded designers, etc. Second, contemporary conservative populism, among other forces, has translated the division into a political one, between suburban, middle American, "real people" and disdainful, unrealistic, urban, coastal elites.

This brief and selective history of the three-plus decades following World War Two shows how the political, economic, and cultural changes in the country served to facilitate and reinforce a particular and particularly American form of suburbia. There were exceptions and variations – historical, regulatory, and regional – that are obscured by the brevity of this account, but conventional suburban development is an overwhelmingly coherent and dominant form. Even this incomplete account demonstrates the manner in which the social and the material converge in the built environment as a social process that both reflects and reproduces the commitments of the context within which it occurs. For a deeper sense of that convergence, we will now turn to the professions caught up in and constituted by these same processes, considering their formation and their practices.

Professional fields and suburban development

We need to move beyond the political, economic, and cultural factors at work in the postwar suburban boom to fully comprehend the social landscape of conventional suburban development. We briefly discussed the emergence and importance of large-scale development operations above and their relevance is ongoing as the economic sector primarily dedicated to profiting from suburbia's construction (they are more comprehensively examined in a variety of other works: see Gans 1967; Hall 1988; Hayden 2003; Jackson 1985; Liell 1952; Weiss 1987). We need also to characterize the states of the professions involved in its conception – its design and planning – and comprehend their roles in the process.

The suburban boom was a planning boom

The explosion of residential, commercial, and roadway development that immediately followed World War Two created similarly radical pressures on the supply of professional planners. In the period between World Wars One and Two, in tandem with the proliferation of land-use regulation, the profession had consolidated its position in charge of the spatial arrangement of the metropolitan built environment. The planning profession responded to the demand for more planners by rushing the formation of new ones. In their haste to fulfill the demand, planners shifted from an apprenticeship model with a practical emphasis on the control of form to an abstract, formal educational model that foregrounded a kind of policy-oriented, numerical rationalism. This shift was apparently resisted by an old guard, but was necessary to produce planners in the numbers called for (Hoch 1994). It is obviously difficult to disentangle the intellectual commitments and shifting theoretical discourse from the demographic pressures on the profession.

Hoch writes, "The newly formed university-based planning schools emphasized the production and dissemination of knowledge rather than the acquisition of practical skills" (Hoch 1994: 47). Practical ignorance in planning was, moreover, justified by Modernist theory, an approach adopted from architecture. Fishman calls "a basic element in the ideology of planning [at the time]: that large-scale public enterprises based on modern design were necessarily superior to the piecemeal process that had built up the fabric of old cities" (2000: 18). As we will see below, this conviction would prove central to planning's ultimate loss of professional autonomy.

The absence of architecture

The planning profession believed that it was integrating key tenets of Modernist theory into its urban renewal efforts, particularly separating functions – residential from commercial and traffic – and combining minimalist design with technology – unadorned towers served by elevators. Architecture was holding still tighter to Modernism and turning inward. Modernism in architecture emphasized a radical break with the past, throwing out historical forms and theoretically putting function foremost, thereby rejecting the vernacular designs to which most homebuyers were accustomed. It also fit with the tendency we have seen in planning by emphasizing the rationalization of urban space – primarily its segregation by use – and by valuing the technological solution – in the metropolis's case, the car and a hierarchy of automobile thoroughfares.

Modernism's dominance in American architectural education and theory endured relatively unchallenged into the 1970s and, though an oppositional, diverse Post-Modernism has since arisen, may still be the most popular, coherent doctrine among several in architectural thought (Larson 1993). In the high Modernist period of American architecture, the profession used its commitment to these principles to reinforce its professional distinction, while drawing away from the broader public and service to it as that public moved to the suburbs. We will see how this tendency created the conditions for its own opposition.

While planners were caught up at the center of the suburban boom, architects were substantially marginalized by suburbanization. The prominence of the highway and the elevation of the automobile can both be understood as profoundly affected by Modernist design theory, but another modern element of the new suburbia – the mass production of homes – effectively excluded architects from the process. Design was subordinated to the dimensions of materials and construction techniques and a handful of designs produced entire new towns (Liell 1952).

This exclusion for economic reasons was compounded by the suburban culture debates: either the suburbs were too low for principled architects to stoop or they rejected architects' presumed aestheticism. The predominance of big builders also exaggerated this problem by introducing an unbridgeable gap between the architect and her ultimate client, the homebuyer. Designing for the middle-class homebuyer in the abstract enforced a common denominator of style that, if not the lowest, was necessarily well below the avant-garde architectural ideal. This tendency toward a broadly acceptable style – embodied in Cape Cod and ranch styles, among others – also defied architecture's attachment to Modernist house designs, since they have never appealed to the broader American public.

This conflict between popular styles and the dominant aesthetic in architecture widened the existing gap between the image within the profession of architect as artist and the reality in real estate development of architect as service provider. Magali Sarfatti Larson, in her sociological study of architectural change in the second half of the twentieth century, finds that "Normal architectural practice is oriented to service and commercial interests, inevitably heteronomous, and often subordinate and alienating" (Larson 1993: 16).

Heteronomy, for Larson, describes the dependence of architects on clients, engineers, and builders to live by designing, to make what they design, and to make it as they design it. Judith Blau writes, " ... owing to the singular master value of design creativity, most architects are destined to fail to realize their aspirations and they know that" (Blau 1984: 59). Thus the dependence associated with heteronomy is not inherently problematic – many professions, after all, serve others – the conflict arises from the juxtaposition of an internally emphasized value and the impossibility of its attainment in the work necessary to gain a livelihood. And the conflict endures because the field continues to reproduce this central value, despite its contradictory character, because it contributes to the field's distinction.

Urban renewal and the failure of modernist planning

While the suburban development of both residential and commercial properties was being actively encouraged in the postwar period – and industry was following it out to the metropolitan periphery – the federal government was also investing in change in urban centers. Comprehensive "urban renewal" was directly legislated in the Housing Acts of 1949 and 1954 (Hall 1988: 227). Carefully kept separate from the actual provision of public housing, urban renewal's primary, if unarticulated, purpose was to eliminate aging, cheap, and often substandard housing and the population that depended upon it from areas that might be profitably redeveloped as commercial space near established downtowns. And this it accomplished: Friedland found that 90 per cent of low-income housing destroyed by urban renewal programs was not replaced (1982: 85). While such an outcome was a logical end of urban "growth machine" politics, the degree of hubris, the magnitude of the intervention, and the mixed reception to and performance of the new spaces would ultimately cause a backlash.

According to Hall, the 1949 and 1954 housing acts that initiated and continued urban renewal "represented another triumph of the NAREB-ULI lobby" (1988: 228), the joined forces of real estate sales professionals (the National Association of Real Estate Boards later became the National Association of Realtors) and developers (the Urban Land Institute). Urban renewal was reformed in the 1965 and 1968 housing acts – displacement of residents was constrained and low-income housing subsidies were mandated – but it remained both on the books and a fiscal loser until the Nixon administration phased out the reformed programs entirely (Logan and Molotch 1987: 169). When urban renewal came to the attention of a public broader than those immediately affected, planners and planning would be tarnished by their seeming complicity with real estate sales and development interests.

The problem with urban renewal was that it did not work, according to Jane Jacobs, at least not in its explicit program of making new places. As an effort to displace the poor from old urban neighborhoods and tear those neighborhoods down, on the other hand, it was quite successful. Jacobs's organization of public opposition to the "Westway" project in Manhattan – championed by perhaps the greatest representative of practical urban Modernization, Robert Moses – and her *Death and Life of Great American Cities* (1961) combined a glowing account of old urbanism, the assertion that urban reconstruction could be resisted, the popular suggestion that traditional urban places could be preserved, and a critique of urban planning. According to Robert Fishman, Jacobs "drastically devalu[ed] theoretical expertise ... returning to the eclectic, pragmatic roots of American planning," making "room for the initiative and creativity of citizens," and, most of all, "She provided a justification for the city" in an era in which it desperately needed one (2000: 19). Jacobs's concepts, like "eyes on the street," offered compelling explanations for the social function of urban form, thereby undermining any planning that failed to take such complex notions into account. Her explanations proved enduring and still hold sway in the criticism of large-scale, subsidized housing developments and the privatization of urban space, among other things (see, for example, Davis 1990).

Jacobs's criticism was so devastating because it highlighted a particular and fundamental weakness of planning in the period: the commitment to rearranging metropolitan space according to abstract principles, rather than practical experience and engagement with the existing built environment. When, by the late 1970s, it had become abundantly clear that the epitome of the Modernist approach to the city in the United States, urban renewal, was a spectacular failure, the planning establishment took most of the blame for ideas that had originated with architects half a century earlier.[4] As Fishman says,

> [urban renewal] combined a poorly understood caricature of modern design with a top-down authoritarianism rarely seen in American planning. Far from saving the cities, urban renewal became a potent symbol for misguided public policy and plunged American planning into a crisis of confidence from which it is only now recovering.
>
> Fishman 2000: 17

Urban renewal had been planning's big chance, and its failure seemed to invalidate both the possibility of large-scale public enterprise and the appropriateness of attempting to improve the urban condition.

Proximate conditions: planning's lack of autonomy

Fishman defines planning, in its original self-conception, as "a collective action for the public good, but particularly action that concentrates on building and shaping the shared physical infrastructure for present needs and future growth" (Fishman 2000: 2) and explains " ... the core belief of the American planning tradition at its most confident [in the 1920s and 1930s]: that intelligent, imaginative, collective action could genuinely shape cities and regions that met the highest ideal of the nation" (Fishman 2000: 14). By contrast, Hoch explains planning's purpose in the early 1990s as "the use of reason and understanding to reduce collective uncertainty about the future" (Hoch 1994: 15). Across the period in question, the field shifted from "building and shaping" to "reducing uncertainty."

But it wasn't just the city and it wasn't just urban renewal that led to the decline of planning. By the 1980s, it was clear that an American suburbia, minimally planned around roadway hierarchies and traffic capacity, was not functioning as expected. The planning field's response to the onslaught of criticism from multiple directions was a retreat from design expertise and the forsaking of normative claims. In its subordination to community oversight, embodied in lengthy and mandated review processes, the profession came to base its authority on the collection of quantitative evidence and the implementation of elaborate land-use regulation, rather than some more qualitative expertise.

Planning practice in local government has become more a matter of explaining plans for public review – to appointed or elected lay commissions and concerned citizens – than one of planning, in the sense of conceiving metropolitan spaces. The professional distinction of the planning field lives in constant conflict with the subordination to the public will. These more general procedural rights guaranteed to the public should not be confused with either consistent responsiveness to public opinion on the part of the planners, or public control of the process. They do make possible the obstruction of development, drawing out permitting and approval to the point that carrying costs may make investment untenable. The implementation and consecration of planning's subordination to public review over the decades since urban renewal has effectively reconsolidated the discrediting the profession faced in urban renewal's immediate aftermath.

Whatever its successes, failures, or actual capacities, planning is nevertheless integrally involved in the contemporary process of suburban development. Conventional suburban development is reproduced in the context of planning's rationalism, its self-imposed ignorance of physical form, and its inability to reach normative conclusions. The current condition of planning must be considered both a constraint and an opportunity because, while the field's members will undoubtedly be as difficult to shift from established practices as any professionals would, a profession in such a state of disarray may be more open to radical innovation than it otherwise would be, particularly if that innovation can provide access to social capital that has not been accessible for decades.

In the absence of an ability on the part of professional planners to positively shape development, anti-developer and anti-sprawl sentiment among local publics has often lead to an unnuanced not-in-my-backyard attitude (or NIMBYism) – finding its ultimate expression in the elegantly named BANANAism ("build absolutely nothing anywhere near anything"). NIMBYism has become an assumed element of the suburban development process and this rather unparticular resistance contests anything that looks like development.

Too many architects, too much theory

If planning overreached and fell flat on its face, we will see that, in the same period, architecture just got in its own way. Metropolitan design and planning concerns were forced outside of architecture by the emphasis within the profession upon creativity and pure design. Suburbia, the dominant American experience of the built environment, can be and continues to be ignored by the most doctrinaire in the profession. Architects as a group thereby work themselves out of relevance to those who would inhabit their buildings. But this does not mean that architects as individuals do not question their own relevance, in light of the readily observable facts on the ground throughout the metropolitan region – suburban housing subdivisions, in particular. The field, like most, has never been without dissenters or quiet holdouts – even at the height of Modernism's dominance, some architecture programs (Yale, Notre Dame) and some notable architects (Robert A.M. Stern) continued more traditional practices.

Architecture's absence from the most dynamic region of development in the postwar period makes all the stranger Robert Gutman's revelation in his late 1980s study of the architectural profession: "Architecture has become the fastest-growing of all the major professions, far outstripping the increases even of lawyers" (Gutman 1988: 24). He goes on to show that this increase in architects is rooted in expanding enrolments and the proliferation of programs in the decades after World War Two. That the numbers of architects exploded concurrently with the profession's lack of participation in the suburban boom has had a profound impact on the conventions of practice for the vast number of newly minted architects. Gutman argues that underemployment, meaning mostly "routine, menial, and low-paying jobs" doing drafting and detail work on large buildings designed by others, particularly for recent cohorts, may serve as something of an apprenticeship and may correct for the increasing academization of degree programs, but also traps many in these jobs indefinitely (Gutman 1988: 28).

By the 1980s, then, many theory-burdened architects were trapped in low-skill work for firms competing ferociously for commissions, even violating the pure art/commercial creation boundary by resorting to developing their own projects. But architecture's professional problems in this period were exacerbated by the aesthetic ones to which we have already referred. Modernist style was popular among corporate clients seeking signature buildings, but never among middle-class homebuyers. Despite the damage to Modernism done by the conspicuous failures of urban renewal and public housing, it is so total an approach that it has proven difficult to replace in the architectural imagination. The subsequent "post-modernism" in architecture is, in essence, too incoherent and reactive to take its place and seems primarily to have increased the disorder in the profession, doing little to address the contradiction between formation and practice while fostering the conditions for a broader and more concerted reaction down the line.

Related to this disaffection and aesthetic separation is the public's increasing distrust of the profession. Gutman writes,

> Users of buildings and urban environments are very suspicious of architects' motives. The profession is looked upon as venal and selfish. Architects are perceived as people who are mainly interested in advancing, often on the basis of spurious arguments, the economic interests of building owners and developers; and therefore, indirectly, the wealth of the professionals themselves. The public's trust in the fidelity of the profession is being undermined.
>
> Gutman 1988: 21

He goes on to explain that the public increasingly has alternatives to depending upon architects for design solutions (Gutman 1988: 62–7) – engineers and construction managers in particular.

The irrelevance of architecture, or at least of the field's definition of merit, is revealed in two empirical observations. First, Larson explains that developers' " ... main criterion in selecting design was and continues to be product differentiation" (Larson 1993: 243). This suggests that the definition of good design internal to the architectural field holds little importance for the purchasers of its work – developers are not looking for good, they are looking for distinct. Second, Blau found that " ... the lower the quality (as measured by awards), the higher the client repeat rate" (Blau 1984: 110). Good design, as defined by the field, may in fact carry negative significance for developers.

Architecture does not have to be irrelevant to suburban development, to the broader homebuying public. Rather the point is that architecture makes itself irrelevant, or at least less relevant, through action within the field related to the reproduction of the field as distinct. When architects organize the commitments or dispositions of the field as they do – emphasizing autonomous artistry and pure design in formation and in discourse – they establish a distance from the general population, from the present needs for design. This distance, shored up by Modernist aesthetics, reinforces their expertise.

As I noted in introducing Bourdieu, fields do not dictate the behavior of their members. Rather, they encourage and enable particular forms of practice. The forcefulness with which the field's boundaries are reproduced is presumably a matter, in part, of that field's ability to reproduce its social capital. The operation of the field does not rule out the opportunity for fully formed members – architects, in this case – to push the boundaries of the field's appropriate practices outward, particularly if they are weakened by shifts in their status vis-à-vis other fields. Or perhaps the boundaries become porous and the disaffected move completely outside the field, but then when their novel practices gain a certain robustness, they effectively demand recognition by the field because of their original formation. However precisely we conceive the metaphor, it should be clear that this practically discredited ideal of architecture opened the possibility of a counter-tendency.

The New Urbanism as the response to opportunity and constraint

Thus did the conditions of suburban development in the United States in the three to four decades following World War Two establish the contours of a complex political, economic, cultural, and professional landscape. The New Urbanism's emergence has depended upon the manner in which it has actively engaged this landscape as one of opportunities to be capitalized upon and constraints to be adapted to. Understood this way, the movement's history makes a new kind of sense, as we shall see in the Chapter 4.

Put slightly differently, the conditions discussed above have become the conditions within which conventional suburban development occurs. Any significant alternative to conventional suburban development must contend with these conditions, because of how they have accreted in the landscape since roughly 1945. They may precipitate crisis in conventional suburban development (opportunity for alternatives), but certainly demand confrontation if any departure from the established way of doing things is to be considered (constraint). We should not presume, moreover, that any of these conditions function solely as opportunities or constraints or that the direction of their influence will be immediately obvious. The private market in housing, for example, makes political intervention difficult, but is simultaneously capable of channeling vast resources into any profitable development activity.

Neither the above nor any of the analysis that follows should suggest that the New Urbanism is or was the only possible alternative to emerge from these conditions. Beginning in the 1970s, a range of responses to conventional suburban development was initiated, from environmentalists' prevention of specific projects or mandating the management of particular impacts, to low-income housing advocates' inclusionary zoning or unit quotas, to preservationists' saving buildings, neighborhoods, or patches of wetland. This has often led simply to the displacement of development – "not in my backyard" becomes "somewhere else," by default, because of the continuing force of for-profit development interests. All of these responses have been piecemeal, addressing some aspect of the impact of conventional development. None has provided a comprehensive, affirmative, self-reproducing model through which to both respond to the complaints against conventional suburbia and to channel development pressure.

By contrast, the New Urbanism seems to have done just that, as the remainder of the book will illustrate. Rather than trying to prevent development, it has lashed up the complaints against suburbia to the realities of established cultural, political, and economic practice. The alternative to conventional suburbia that New Urbanism and New Urbanist places embody, we will see, is an integrated and comprehensive one. Importantly, it is neither a revolution nor a rejection of suburban development, but rather a pragmatic realignment. What the following chapters reveal is a movement that has recognized the necessity of functioning within a system that provides housing through a private market and thus knows it can only go as far as it proves itself profitable. This is the foremost of the constraints it faces.

But we know the story is not simply one of constraints. We will also see that the New Urbanism seizes as an opportunity the uniformity and homogeneity of contemporary suburbia and the segregation of uses. Before proceeding to the analysis of the movement's emergence in terms of opportunity and constraint, the subject of Chapter 4, we need a more detailed introduction to the New Urbanism. Chapter 3 will accomplish that by reconstructing the movement's origins through the words of its founders. We will also consider key discursive mechanisms and practices integral to the effort and examine an early New Urbanist development, the Kentlands in Gaithersburg, Maryland, recognizing its relevance for the reproduction of the burgeoning social-material phenomenon.

3 Making a movement

The New Urbanism began as an informal network of architects with a shared educational background. This chapter begins with the movement's founders recounting its origins in their own words, revealing who came together, how, and why various interpersonal dynamics became important for the ultimate shape of the New Urbanism. This core group both organized a "congress" – that has grown into a 3,000-plus member collectivity; wrote a charter that articulated and defined the movement's commitments; and developed the "charrette," a cooperative planning technique that reproduces the movement. At the same time, progress in the organizational and intellectual realms could not have been made without the literal grounding of these efforts in early developments like the Kentlands, in Gaithersburg, Maryland. While the places depended on the group's ongoing development of theories and their cooperation, attracting new members to the movement required built examples to prove the principles and their potential.

From idea to organization

Various interviewees among the founders and early adopters suggest that the core group of New Urbanists was originally formed from two informal circles on separate coasts that first came together in the early 1990s. The West Coast circle included founders Peter Calthorpe and Daniel Solomon and early adopter/internal critic/fellow traveler Douglas Kelbaugh. Calthorpe and Kelbaugh knew each other as undergraduates in architecture at Yale, then ran a firm together, starting in 1989. Calthorpe and Solomon both taught at UC Berkeley's School of Environmental Design. The West Coast circle was united by a common interest in low environmental impact home design, particularly the passive solar house. But Calthorpe and Susan Benson researched and wrote "The Solar Shadow" in 1979, showing that a narrow focus on conventional solar solutions – such as rooftop solar panels to generate electricity for home use – obscured greater energy consumption issues, like the common necessity of automobile use. After that, the circle's interests shifted toward regional design or the spatial arrangement of built elements and their uses within the region, in particular, the relationship of segregated land uses and transportation needs.

Most of the East Coast circle, primarily Andres Duany and Elizabeth Plater-Zyberk, Stefanos Polyzoides and Elizabeth Moule, met as undergraduates at Princeton University.[1] Both couples are married and are the principles in firms that bear their names – commonly known as DPZ and Moule and Polyzoides, respectively. By the late 1980s, Polyzoides and Moule were actually based in the Los Angeles area and, thus, were not geographically East Coasters, but fall in that circle as a matter of original social ties. Moule says:

> In Los Angeles, we didn't find a lot of colleagues that shared our views. Our friends were Frank Gehry and Thom Mayne, and others. The architectural work being done in California at the time, and still being done in Los Angeles to a great extent, has more to do with the development of a single object building than it does with the entire fabric of a place. So instead of doing the usual thing, which was joining up with the local AIA [American Institute of Architects], we decided to seek out our old classmates and friends across the country, that we had more in common with, to see what information we might be able to share with one another. We had always been very good friends with Andres and Lizz and, for that matter, they're one of my children's godparents.
>
> Interview with author, January 6, 2006

Douglas Kelbaugh functioned early on as a facilitating link between the West and East Coast groups, stemming primarily from his connection to Polyzoides at architecture school – though he was also doing his graduate work at Princeton while Duany and Plater-Zyberk were undergraduates: "I was kind of the one guy that knew everybody" (Interview with author, June 10, 2005 [hereafter "Kelbaugh"]).

Early on, this core group was enabled, expanded, and reinforced by connections through academic lecturing. Bridging the east-west divide, Duany says, "And I knew about Calthorpe because – essentially the lecture circuit brought us together. The lecture circuit is of incredible value – I don't know how much influence the lecturer has on an audience, but the lecturers certainly get to know each other" (Interview with author, November 8, 2005 [hereafter "Duany"]). Indicating some influence on the audience, in 1980, Dhiru Thadani heard Duany speak on Finnish Modernist Alvar Aalto at Catholic University where he was teaching:

> And a year later I was on the lecture committee … and I said, "We've got to get this guy back to give this lecture on Aalto, because everyone's got to hear this." So we called him up and said, "Hey come on down and give this lecture." And so he accepted … . Then, somewhere along the line, he called up and said, "Listen, I'm kind of past giving that lecture, it's really hard to, you know, all the pronunciations of all the places, I've kind of lost track of it. But I've got this kind of idea, something about small towns in America."
>
> Interview with author, November 11, 2005 [hereafter "Thadani"]

Duany then spoke at Catholic about early ideas for traditional town planning, especially preliminary work on Seaside, the small, traditional resort town on the Florida panhandle which we visited in Chapter 1, and drew Thadani into a Miami-based circle. Ultimately, one of Thadani's classes at Catholic helped Duany and DPZ with the code for Seaside and built a model of the town that served as a major marketing tool (Thadani). The national attention that Seaside garnered for Duany and Plater-Zyberk in the early 1980s was perhaps the most important mechanism for putting together a functional group of like-minded designers.

Duany and Plater-Zyberk also forged links to a number of south Florida architects through their Miami-based firms – first Arquitectonica and then DPZ – and their work at the University of Miami – colleagues Victor Dover, Joseph Kohl, Jaime Correa, and Erick Valle were all important early in the movement. However, the early opportunities for New Urbanists in academia should not be overstated: Stefanos Polyzoides taught at the University of Southern California for many years before resigning in frustration, "citing ideological incompatibility."[2]

Beyond social ties, it is essential to understand the founding group as uniting a set of disparate substantive issues of design. Solomon says:

> … there were half a dozen people, but we knew of a somewhat larger group of colleagues who had similar recognition of the same disintegrating forces of American cities, who were struggling against that in a variety of professional roles with enormous frustration and enormous lack of collegiality. It was an opportunistic banding together of people of similar values, similar sorts of frustrations.
>
> Interview with author, June 11, 2005 [hereafter "Solomon"]

Social and professional connections enabled the amalgamation of multiple concerns to produce a holistic vision for the reform of the entire metropolitan region. The founders fell into loose categories of West Coast-environmentalist and East Coast-traditionalist, but can also be thought of according to their respective interests in the scales by which they divide the larger New Urbanist concept in the Charter: building, block, and street; neighborhood, district, and corridor; region.

Moule and Polyzoides were most conspicuously concerned with the smallest scale, concentrating in particular on the energy efficiency and environmental benefits of vernacular styles in the Southwest. They were also drawn to the problems of multi-family housing and building massing that affect users' perception of the street as an "outdoor room." Similarly, Solomon's greatest interest is probably in building style; his designs are notably Modernist. He has been called the most talented designer of the lot. While this focus ties him to Moule and Polyzoides, he also espouses a conception of an urban system interconnected at every scale:

> [a] model of the organization of the metropolis in its physical structure, and economic and political structure, into which a whole mosaic of things fit and to which they contribute. But it's a big frame, a big model, around which highly particularized things are organized.
>
> Solomon

Duany and Plater-Zyberk were most active at the neighborhood scale. Much of their early work concentrated on the analysis of successful, existing, traditional neighborhoods and on the ideal spatial arrangement and interrelation of neighborhood functions. They have also been most concerned to develop a regionally appropriate, traditional aesthetic. Calthorpe, in contrast, was most focused on the region. He has, to a greater degree than the others, emphasized public transportation and "transit-oriented design" as integrating the region. More recently, Calthorpe has published an innovative thoroughfare design for dense downtowns.

What all of the founders shared was the belief that any solution would depend on operating at all scales; their actual work ultimately defies this early substantive categorization. They all work at more than one scale, both because of the generalism of New Urbanist practice and the varying scale of development. Nevertheless, their writings and public positions tend to conform to this early categorization.

Other concerns also animated the founders and took on new significance over the course of the movement's development. Moule and Polyzoides, in particular, have jumped the original east-west divide to become the most conspicuous practitioners of green design. Their conspicuously green work runs from an early neighborhood-scale project in Arizona, Civano, to the award-winning Robert Redford Building for the NRDC in Santa Monica, to a recent, large subdivision outside of Las Vegas. They have implemented comprehensive environmental impact evaluative practices for their firm's projects and emphasized energy efficiency and durability in choices of building materials. Solomon's interest in and championing of Modern design at the scale of the building continues to separate him both from those relatively unconcerned with design at that scale, like regionalist Calthorpe, and from those who emphasize traditional design and historical precedent, notably Duany, but also Polyzoides, who stresses local vernacular. Solomon's continuous critique of the dominant New Urbanist approach to style, and his resistance to traditionalism in particular, has kept current a discussion within the movement that would otherwise likely have dissipated and has also made space in it for younger designers interested in stylistic issues and turned off by traditional designs.

In sum, a complex network of personal ties, and the ample social capital it encompassed, reinforced connections among the core group across a decade of informal interactions during the 1980s. The group was driven forward as a collective by the nature of their substantive concerns, the manner in which these transcended original social ties, and the productive tension among them. The question of how this group transformed itself into a movement and built an organization to underpin that movement, in the most concrete terms, is our next concern.

In the late 1980s, Peter Katz, a graphic designer and marketer, met Polyzoides and Solomon, among others, in Seattle while on hiatus from his marketing career. They met at a yearly event at the University of Washington organized by Kelbaugh, who chaired the architecture department there from 1985 until he was named Dean of the Taubman College of Architecture and Urban Planning at the University of Michigan in 1995. Katz had been exposed to Duany and Plater-Zyberk's early work on Seaside when they presented at a workshop organized by Peter Eisenman in New York in 1983. He remembers being initially scornful: "And I just thought that Seaside was the stupidest thing in the world. Why are we building this little New England fishing village on the Gulf Coast? What lessons could we possibly learn about architecture and urbanism?" (Interview with author, November 29, 2005 [hereafter "Katz"]).

Through the Seattle event and his introduction to Polyzoides and Solomon, Katz came to understand that Duany, Plater-Zyberk, and like-minded designers around the country were trying to actually build places that captured the values on which he was relying in his marketing work for big builder clients. He talks about trying to sell conventional suburban projects – single-use and car-dependent – by just calling them "-walk" or "-village," without addressing their actual form. His exposure to the gathering, proto-New Urbanism forced him to conclude, "Why lie when you could actually build it?" (Katz).

The Seattle event that got Katz involved was referred to as the "Pedestrian Pocket" charrette. This charrette was an early version of what became a central New Urbanist design method. The charrette is an established practice in architectural education: a collaborative, critical, high-intensity, problem-solving session addressing a particular design issue, often for a deadline. This collaborative approach values everyone's contributions to the resolution of particular design problems, opening opportunities for early career professionals and lowering the barriers to speaking across disciplines (Brain 2005b). As we will see, it has been reconceived by New Urbanists, in this case as periodic meetings of a group attempting to refine the pedestrian pocket concept, a transit-oriented, neighborhood design that originated with Calthorpe and grew out of his post-solar work. It tried to integrate neighborhood functions at manageable walking distances from transit hubs. This series of annual charrettes resulted in the technical *Pedestrian Pocket Book* (Kelbaugh 1996), a bestseller in urban design, according to Kelbaugh, and probably the first book-length treatment of a New Urbanist-related or cognate subject.

Katz explains that, having met Polyzoides at the charrettes, he was subsequently persuaded to write a book on the earliest built efforts of the group of founders and some of their associates. The book (Katz 1994) would be about them, but also for them, consecrating the movement's efforts. This would not be Katz's first authorial experience – he had earlier published a work on Richard Meier, a renowned Modernist architect – and he conceived of himself as a kind of journalist in addition to his other occupations.

> I showed Stef [Polyzoides] what I did as a marketer and Stef got extremely pumped and said, "Here's a willing body that we can get to do this book." It was like herding cats ... it was this weird project that seemed kind of propelled by a certain destiny, there were countless times where it was dead and then it would somehow revive and move forward.
>
> Katz

The events that followed eventually resulted in the formal foundation of the movement, but arose from the confluence of the founders' interest in publicizing their efforts and building their practices with Katz's interest in both completing and selling his book project. This is not to deny an underlying substantive motivation on anyone's part, merely to claim that the foundation of the movement played out the way it did because of these particular conditions.

By fits and starts, Katz became the marketing and organizational presence in the early stages of the movement. Katz had the expertise to publicize their work and time the architects could not afford without compromising their ongoing design efforts:

> I basically came along and I said to these people, "There's a movement here. You're all so busy fighting with each other about the minutiae, but the fact of the matter is that you have much more in common than you have different and there's a great value here in coming together as a group and defining yourself as a group. It will be transformative, both in terms of your practices as well as the state of American urbanism." And that was my mantra, that was the thing I kept saying over and over again. And they listened. And basically they allowed me to apply my marketing skills to their thoughts.
>
> Katz

Katz explains his role throughout as both impassioned and pragmatic, conceptual and organizational, out front and behind the scenes:

> Now there's an interesting irony … since they resent that anyone remotely thinks that a marketer is connected to the movement. They believe that this movement happened because of their brilliance, their vision alone, and the fact that I'm in the margins erodes the myth, in their view. To them it was immaculate conception. But that can be our little secret.
>
> Katz

It was a relationship not without its tensions. Katz says that, because he was not an architect, and the New Urbanism is a design-based movement, his public and conceptual role has always been minimized. Still, it is obvious that the course the movement took was dependent in part on his specific abilities.

Two key events took place in 1991, both of which played a part in the movement's taking form. First the name, "the New Urbanism," was coined in conversation between Katz and Polyzoides during a 1991 planning session for an early cooperation between Moule and Polyzoides's and Duany and Plater-Zyberk's firms, Playa Vista. Katz says:

> Stef had been trying to talk me into doing this book and I said, "As a marketer, I think this movement needs a better name." And I suggested a few of my suggestions, he suggested his, and one of his was "the New Urbanism." I said I thought that was a good choice for a number of reasons.
>
> Katz

The name branded the movement, encapsulating it and making it easier to present both as novel and as a coherent phenomenon with a comprehensible referent, something Katz and Polyzoides agreed it needed if it were to attract significant attention. Katz, moreover, needed an appealing name to sell the unfinished book project to a publisher.

Second, at Peter Katz's instigation, the Local Government Commission (LGC), a non-profit that "assists local governments in establishing and nurturing the key elements of livable communities,"[3] sponsored a meeting in Sacramento, California (Corbett and Velasquez 2004). LGC gathered what would be the founding group – except Solomon – along with Michael Corbett and asked them to articulate a set of principles for the design of community and region that LGC could disseminate as recommendations. The meeting, proverbially culminating over a pasta dinner at LGC Executive Director Judith Corbett's house, was important for two reasons. First, while the LGC had called the meeting for its own purposes, it put most of the founders of the New Urbanism together as a group for the first time for a collective purpose beyond the practice of design. In particular, the LGC's request pushed them to articulate their critique of contemporary development for a broad public and to state a positive direction for a potential alternative. Second, it produced a document – the Ahwahnee Principles (presented at the Ahwahnee Hotel in Yosemite National Park to a conference of mayors and elected municipal officials) – that would form the basis for the Charter of the New Urbanism. As we will see later in a discussion of the Charter, the articulation and establishment of a set of principles was a central mechanism for the reproduction of the movement.

Katz brought the founders together again in 1992 for dinner at the Lotos Club in New York. This was an effort to foster the cooperation necessary to finish his book and also to begin planning a related conference on the group's work. The founders had been discussing a congress, amongst themselves and with others, and Katz hoped such an event would help publicize his project, "I liked the idea of a conference, because I thought it would help legitimize the book" (Katz). According to Katz, the assembled group quickly vetted his current draft manuscript – "They just wanted to know that I wasn't going to screw them or rip them off in some way" – and moved on to a conversation about naming in which "the New Urbanism" was rejected and replaced with the "Congress for Urban Reform" (CUR), which Katz would not be allowed to use as the title for his book. Katz says:

> Meanwhile, I had already sold the publisher on the term "New Urbanism," I was in contract on it, and so when they decided to call it the "Congress for Urban Reform," I went fine, because as a journalist, I'm going to decide to call the book what I feel is the name of the movement.

The New Urbanism: Toward an Architecture of Community came out in 1994 (Katz). This would be the next significant event in the movement's development. *The New Urbanism* is less a book "about," written from the outside looking on, than it is a book "of," a catalog of project descriptions, position statements or explanatory musings, and images contributed by the founding practitioners and early adopters. Katz claims that the movement's name was ultimately reconsidered because the book's title was set and it was forthcoming as of the first congress:

> And then the funny thing is about a few months before the CNU [first congress], Andres realized that my book was coming out shortly and that my book was going to be the thing defining the movement and so it was just stupid not to [use the name].
>
> Katz

The opportunities for cross-marketing were too great for Duany, at least, to ignore – Katz calls him an instinctive marketer. The congress's title and, thus, the movement's, were apparently established by Duany's fiat.

We should note in passing that the choice between the "Congress for the New Urbanism" and the "Congress for Urban Reform" and, in particular, its resolution, is reflective of a sensitivity and pragmatism that would prove integral to the movement's success. The choice of "New Urbanism" over "Urban Reform" had the virtue of suggesting a novel direction. It was also an act of positive reclamation rather an oppositional position-taking with problematic resonance – "urban reform" is reminiscent of "urban renewal". There is no suggestion from among my interviewees that name "the New Urbanism" misrepresented some essential character of the budding movement, but it did shift its public first impression away from confrontation. Once this contest was resolved in the run-up to the first congress, none of my informants suggest that the decision was ever second-guessed.

After Katz's book, the congresses themselves were the next significant step in the movement's formation. The first three congresses were planned around the three scales at which the New Urbanist concept was intended to operate – building, block, street; neighborhood, district, corridor; region – and each was organized by the founders identified above with those scales – Moule and Polyzoides; Duany and Plater-Zyberk; Calthorpe and Solomon, respectively, though not in that order. The point of the early congresses was the presentation and critical evaluation of projects in the interest of sharing information and experience and refining concepts, thus meetings were focused on a quick succession of presentations by designers in marathon sessions emceed by Duany:

> everyone had a little clip of information that [Andres] knew they knew and he would say, you're going to come talk about that, and so it was loaded with stuff. All 200 people got to say something, but it was just for fifteen minutes, really a high learning curve.
>
> Thadani

In 1993, the first Congress for the New Urbanism, "CNU I: The Neighborhood, District, and Corridor" was held in Alexandria, Virginia. Dhiru Thadani, based in Washington, DC, facilitated its site selection. He says that he pushed for Alexandria with Duany and Plater-Zyberk both because Alexandria included some of the best traditional urbanism on the East Coast and because of his connection to then Dean of architecture at Virginia Tech, Gregory Hunt. Thadani says of Virginia Tech that it "was totally uninterested, as a school, in any of this stuff. But I knew Greg and the facilities were there [at the College of Architecture's Alexandria campus]" (Thadani).

Attendance at the first three congresses was by invitation only, "if you knew Andres and he knew you and you didn't do suburban work, you were invited. Really small group, about 200 people" (Thadani). The foregoing suggests the primary origins of these various connections were educational and professional. "CNU II: Buildings, Blocks, and Streets" was held the in 1994 in Los Angeles and was hosted by Moule and Polyzoides. Then "CNU III: Nature, Infrastructure, and the Region" was held in San Francisco in 1995 and was hosted by Calthorpe and Solomon. CNUs II and III followed the same format as the first meeting. Anecdotal evidence suggests that gaining an invitation if one was interested was not difficult.

All of my interviewees who attended the early congresses, both those who remain important in the movement and those who have stepped back from it, look back fondly upon them as moments of great intellectual ferment. Ferment apparently included much friction and disagreement, mostly channeled productively by attendees. No doubt it helped that most were architects, thus already familiar with the abrasive, architectural-educational evaluative form of the "crit," the review by faculty and peers of projects standard in architecture school: "We don't do this so much anymore, but the process of 'critting' each other was incredibly valuable" (Raymond Gindroz, interview with author, June 9, 2005).

Between 1993 and 1995, the movement's founders exhausted the pre-established themes for the congress – the three scales – and their own organizational energies. Early discussions had envisioned the congress as a limited intervention, resulting perhaps in some published proceedings, then disbanding. There is no suggestion that any of the founders originally imagined an ongoing formal organization. In fact, Solomon describes the movement as "a vigorous evolution of energetic initiatives that never had a plan" and says, "There are evolving strategies of how to respond to opportunities and nurture them, but it's not like the Communist Party or the Catholic Church, it's not something that has evolved from doctrine with a plan for world domination."

Four factors helped to sustain the movement, continue the annual congresses, and formalize an organization despite this exhaustion. The first was the combination of Katz's willingness to expand his role as an organizational force for the movement and his identification with it because of his authorship of *The New Urbanism*. The second factor was the financial support of not-for-profit funders for the fledgling effort. Katz explains this factor in detail:

> A lot of the foundations had been funding a lot of land conservation stuff and inner city stuff in the hopes of influencing legislation. And when the Contract for America blew in, the foundations were just disillusioned, and they went, "Man, we've been spending all this money and all these people we got elected are gone!" And they looked around and they said, "What's working?" And the [Adler 1995] *Newsweek* article [about the New Urbanism]

> had come out and they said, "Who did that? How did that happen?" And they learned that it [the article] was a result of my efforts and I got a phone call from the Energy Foundation, which was a consortium of Rockefeller, Pew, and Macarthur, and they said, " … we'd like to fund your outreach efforts."
>
> Katz

The funders also contacted the movement's leaders and a lengthy struggle between Katz and the founding architects ensued over the precise definition of Katz's role in what would eventually become the non-profit Congress for the New Urbanism. Because he was identified with the effort to draw attention to the movement and because his work had helped attract the funding, Katz had to be involved, despite the fact he was not an architect. Moreover, the funders demanded a degree of formal organization for which Katz was prepared, but which conflicted with the founding architects' sense of a movement driven by principle – and would have conflicted with their design practices had any of them taken it on. Ultimately, Katz was named *Acting* Executive Director:

> So suddenly, the Energy Foundation gave us a $50,000 start up, Macarthur gave us another $50,000, and the thing was running out of my apartment in San Francisco … . My board didn't give me a lot of freedom … but I did some interesting policy stuff and most of it was done without their knowledge … . But they'd show up for the glory and that was the unwritten rule, I would do the grunt work behind the scenes.
>
> Katz

This establishment of a formal organization to serve the movement is the third sustaining factor, because it concretized a sense of collective, political purpose.

Katz toured the US and Australia speaking about the movement on the CNU's behalf, generating speaking fees that he shared with the organization, but most of his early responsibilities were around CNU IV which took place in Charleston, South Carolina, in 1996. A shift from the first three congresses to a new format for the fourth and beyond is a fourth and final sustaining factor for the movement. Two things about CNU IV are particularly important and are causally connected, according to Duany. First, the congress was opened to the public. Second, the nearly 300 attendees were presented with the Charter of the New Urbanism and offered the opportunity to sign onto it – thus producing the distinction "charter member".

The founders were only content to open membership when they had established some clear definition of membership in what. Duany says, "An interesting thing happened when the Charter was written, which is we felt we could open the membership" (Duany). While the movement was something like an ongoing, critical conversation among tens of generally like-minded professionals, its organizing assumptions could remain implicit or only loosely defined, and the theoretical and technical content of the conversation could go on without unwieldy explanation. This informality, however, could not accommodate much growth in the number of participants, while preserving a sense of mutual agreement or, at least, understanding. The unified and unifying statement of principles that the Charter embodied defined the movement's terms and established a baseline for discourse within it, serving, thus, both as a boundary drawing and a disciplining mechanism: "Keeping the discourse at a high level, you basically lose the people who want a more simple answer," (Duany) they fall or are pushed outside a clearly drawn boundary. By stating outright what the New Urbanism was, the Charter made explicit the terms of allegiance to the movement and made possible the identification of violation of the movement's principles as a basis for exclusion from it, effectively the terms for membership.

Harder to account for, in these early years, is the role of charisma. Andres Duany, most conspicuously, has served from the start as a charismatic leader for the movement. Some of my non-architect interviewees have referred to him as a "Pied Piper" and have suggested that people would have and did travel distances to hear him speak repeatedly. Others have suggested that for the movement's sometimes beleaguered general membership, he appears a champion. Duany remains a symbolic head of the movement, but also a lightning rod. Yale architecture professor, Alan Plattus, says of Duany:

> Andres ... as a friend and colleague, he can be absolutely fantastic. He's smart, he's witty, he's a thoughtful, rational person. As a public figure these days, I think he can be infuriating. Demagogic, dogmatic. I understand that he feels that strategically one needs to do that sort of thing.
>
> Interview with author, November 15, 2005

Duany admits this role and seems to embrace it:

> ... my particular role is keeping us from becoming certain, almost exactly the opposite of what people think I do ... . Basically, I have the kind of practice that, I have too much work, so basically, I don't lose work for saying unpleasant things. Some of us don't have that privilege. I use that privilege to bring up things that would cost me work.
>
> Duany

To architects who value individual genius, any sense of being led can rankle. Kelbaugh describes Duany as wanting to be "the commander" and says of other architects' reactions to him:

> It was originally a broader pool ... some people think Andres is just too dominating, and people either like or don't like Andres. He's gained, but I think he's also lost a lot. I think a lot of people have dropped out because the CNU is more or less his personal thing.
>
> Kelbaugh

Kelbaugh does not suggest Duany is the only significant personality in the movement; in fact he insists that he never could have gotten as far as he did without his wife and co-principle, Plater-Zyberk, to smooth ruffled feathers:

"Lizz is a saint, I mean she really is amazing ... like a member of the royal family or something, the good part of that" (Kelbaugh). Moreover, Calthorpe can be as dominating and "wants to be influential, but he's not interested in running the CNU at all" (Kelbaugh). Some of the CNU's relative ideological neutrality, in fact, must derive from the balance between Duany's conservatism and Calthorpe's wry, left-coast liberalism.

Another critical development for the movement was its alliance with the federal government. In 1996, the CNU, and Peter Calthorpe's firm, in particular, worked with the Department of Housing and Urban Development (HUD) on guidelines for the redevelopment of federally subsidized, low-income housing, a program that became known as HOPE VI. Calthorpe's left-leaning politics were the right ones for that moment. Many New Urbanists are interested both in so-called "urban infill" and in low-income housing. HOPE VI offered an opportunity to pursue those more urban and progressive interests, which are often neglected in favor of more conspicuous and marketable greenfield projects. The cooperation with HUD substantially raised the profile of the CNU, both as a movement and as an organization. One of the Charter signatories and probably the most distinguished attendee of CNU IV – among a number of mayors and academics – was US Secretary of Housing and Urban Development Henry Cisneros. This cooperation and Cisneros's attendance also generated at least a loose association with the Clinton Administration that would become more important with Vice President Al Gore's presidential campaign in 2000, as Gore focused on smart growth and sustainability. We will consider the New Urbanism's relationship with the federal government in more detail in Chapter 5.

As the annual congress became a larger, more professionally, and more routinely produced event, two important landmarks followed for the organization. Both of these were changes in leadership of the non-profit that consolidated the movement's activities and underpinned it reproduction. First, in July 1997, Katz stepped down from running the CNU and Shelley Poticha became Executive Director. While the CNU remained based in San Francisco, Poticha's tenure has been described as the period in which the New Urbanism was transformed from a "boutique movement into a mass movement."[4]

Katz suggests that Poticha, trained as a planner and coming to the CNU from Calthorpe's San Francisco firm, brought to the job both experience in policy-making and Washington lobbying and a West Coast, regional and environmental policy bent. He claims that Poticha increased the viability of the CNU as a grantee, while using the same emphasis to expand the membership beyond architects to planners and municipal officials more concerned with policy issues, perhaps somewhat to the detriment of the focus on design. Accordingly, this was not a uniformly approved of direction: one of the founders referred to it as the "Berkeley episode, very left wing, and we got actually saddled up with Smart Growth and the Republicans came out after us. And we're still recovering from the Berkeley episode."[5]

In 2004, Poticha moved on to lead Reconnecting America, an organization championing transit-oriented development, whereupon the CNU moved its offices to Chicago and hired John Norquist, longtime Milwaukee mayor and

CNU participant, as President and CEO. This second change in leadership was part of the "recovery" from the "Berkeley episode"; the geographic shift of the headquarters from the West Coast to the middle of the country was decidedly symbolic. The same founder cited above about the "Berkeley episode" says about this change, "So basically, we moved to the Midwest, to Chicago, things are now 'righted' a lot better" – though this person also goes on to admit that the post-Katrina association with conservative Mississippi governor Barbour risks pulling the CNU too far right.[6]

Norquist has been primarily tasked with growing CNU membership and increasing the organization's influence. Under his leadership there has been a significant increase in staff – from one in 1995 to seven, currently, including directors of planning and communications and coordinators for the congresses, membership, and research. He has been trying to grow the group through "energizing" topic specific task forces and committees – education, low-income housing, etc. – and expanding the number of CNU state or regional chapters – of which there are currently twenty-one in various stages of formation. Local chapters are a recent organizational innovation that the CNU hopes will generate a new level of member integration through ongoing local activities in addition to the annual national congress. Norquist has undertaken these tasks while trying to preserve the character of the CNU. He said: " ... we're not trying to compete with membership organizations like ULI or AIA. ... We're a movement, we're not an occupational licensing – we're not barbers" (Interview with author, June 9, 2005 [hereafter "Norquist"]).

Most important in the recent expansion of the movement, or at least its public profile, has been the organization's response to the devastation of the Gulf Coast by Hurricanes Katrina and Rita in the fall of 2005. Within weeks of the hurricane's making landfall, the CNU assembled a hundred-plus person team of designers, planners, and specialists and sent them to Biloxi for a charrette with a roughly equal number of Mississippi professionals to plan the reconstruction of the Mississippi Coast. No other organization responded as directly. The ability to respond to the crisis in this manner displays both the loyalty to the organization felt by its members and the CNU's organizational capacity.

The foregoing portrait shows the transformation of the movement from a loose set of aesthetic and personal associations into a professionally managed, non-profit organization with 3,000 members worldwide and an annual congress that attracts approximately 1,400 attendees. This involved the building and consolidation of social capital in a founding group; the enlistment and service of dedicated and variously skilled organizational leaders; and the attraction of attention and financial support on a sufficient scale to enable expansion and reproduction. We should note how radically contingent each of these elements was: many were based in social ties arising from educational formation, but their combination and persistence was effected and affected by everything from financial pressures on a single individual – Katz struggling to get paid for his book – to national-scale, political-economic shifts – the Contract with America and the foundations' response. Referring back to Molotch's concept of lash-up, the New Urbanism could only become precisely the phenomenon that it has through the precise synergy of contingent factors, both structural and agentic.

The Charter as discursive tool and the charrette as practice

The Charter of the New Urbanism is mentioned above, but deserves further attention from a sociological perspective as a key mechanism for the New Urbanist efforts to build a movement. More than just a historical marker of the movement's foundation, the Charter defines membership and lends New Urbanists a kind of leverage in debates about the substance of their work. Relatedly, much New Urbanist design is done using a method referred to as the "charrette." The charrette brings design and planning specialists together with local stakeholders to brainstorm "solutions" to the design problem that a new development presents. But we will see it does much more than that, in constituting a core practice for the movement and its members.

The Charter of the New Urbanism

As noted above, in 1996, at the fourth Congress for the New Urbanism in Charleston, South Carolina, the Charter of the New Urbanism was ratified by the 266 attendees, including HUD Secretary Henry Cisneros.[7] The preamble identifies "disinvestment in central cities, the spread of placeless sprawl, increasing separation by race and income, environmental deterioration, loss of agricultural lands and wilderness, and the erosion of society's built heritage as one interrelated community-building challenge" (Leccese and McCormick 2000). It goes on to explain that, though the undersigned do not believe the solutions to these problems are as simple as improvements in policy and design, "neither can economic vitality, community stability, and environmental health be sustained without a coherent and supportive physical framework" (ibid.).

The body of the Charter is made up of twenty-seven claims about the appropriate motivations for design and the social purposes of the physical form of the city, broadly understood. The claims address three scales, with nine parallel claims at each scale, in descending order of magnitude: the region; the neighborhood, district, and corridor (functionally distinct); and the block, street, and building.[8] It was apparently "written in committee, but sequentially, a few people writing then handing over to others, then handing over to wordsmiths, then handing over to others" (Duany).

They start by defining terms to clarify what is important to them and move on to recommendations. Some examples are shown opposite.

Kelbaugh suggests the Charter represents the best side of a somewhat elite movement: "It is a sort of upper-middle class movement, but I think there's some genuine idealism. I mean, the Charter is pretty hard to fault" (Kelbaugh). It transcends a class-based set of interests to broadcast progressive claims with seemingly universal appeal. Various other informants have suggested that there is little in it with which anyone could disagree – both a virtue and a weakness.

Duany says of the Charter:

> It's absolutely fucking brilliant. It has held up incredibly. It answers all the critiques and all the questions. When people say, "What about this and what

The region: metropolis, city, and town

1 Metropolitan regions are finite places with geographic boundaries derived from topography, watersheds, coastlines, farmlands, regional parks, and river basins. The metropolis is made of multiple centers that are cities, towns, and villages, each with its own identifiable center and edges.

2 The metropolitan region is a fundamental economic unit of the contemporary world. Governmental cooperation, public policy, physical planning, and economic strategies must reflect this new reality.

3 The metropolis has a necessary and fragile relationship to its agrarian hinterland and natural landscapes. The relationship is environmental, economic, and cultural. Farmland and nature are as important to the metropolis as the garden is to the house.

4 Development patterns should not blur or eradicate the edges of the metropolis. Infill development within existing urban areas conserves environmental resources, economic investment, and social fabric, while reclaiming marginal and abandoned areas. Metropolitan regions should develop strategies to encourage such infill development over peripheral expansion.

The neighborhood, the district, and the corridor

10 The neighborhood, the district, and the corridor are the essential elements of development and redevelopment in the metropolis. They form identifiable areas that encourage citizens to take responsibility for their maintenance and evolution.

11 Neighborhoods should be compact, pedestrian friendly, and mixed-use. Districts generally emphasize a special single use, and should follow the principles of neighborhood design when possible. Corridors are regional connectors of neighborhoods and districts; they range from boulevards and rail lines to rivers and parkways.

12 Many activities of daily living should occur within walking distance, allowing independence to those who do not drive, especially the elderly and the young. Interconnected networks of streets should be designed to encourage walking, reduce the number and length of automobile trips, and conserve energy.

13 Within neighborhoods, a broad range of housing types and price levels can bring people of diverse ages, races, and incomes into daily interaction, strengthening the personal and civic bonds essential to an authentic community.

The block, the street, and the building

19 A primary task of all urban architecture and landscape design is the physical definition of streets and public spaces as places of shared use.

20 Individual architectural projects should be seamlessly linked to their surroundings. This issue transcends style.

21 The revitalization of urban places depends on safety and security. The design of streets and buildings should reinforce safe environments, but not at the expense of accessibility and openness.

22 In the contemporary metropolis, development must adequately accommodate automobiles. It should do so in ways that respect the pedestrian and the form of public space.[9]

> about that?" You say, "Read Chapter 19 of the charter." It really is quite an incredible instrument. It shows you what you can do organically with people who've actually been through the same experience.

Duany reveals how the Charter both unifies movement participants through their discursive application of a common resource and reinforces the movement's discursive strength by providing an authoritative reference. William Lennertz, Executive Director of the National Charrette Institute, a Charter signatory and long-time New Urbanist, echoes Duany when he explains that the Charter articulates and preserves the movement's mission and, by laying out the movement's core values, provides something coherent around which the membership can come together (Interview with author, December 2, 2005 [hereafter "Lennertz"]). It is easy to see why the current CNU President and CEO, John Norquist, calls it "probably the most significant accomplishment of the CNU so far" (Norquist).

Moreover, the Charter, and the values it embodies, works both as one of the movement's primary recruiting tools and as a disciplining instrument, because the founders and early adopters assembled in it the mechanism necessary to keep a broader membership in line. Norquist says:

> The Charter's a pretty persuasive document. I mean, it's not like we're going out and begging people to join. Growing membership is good but we're not looking to have just anybody walk in the door, we want them to understand the principles and be committed to the cause.

By articulating the movement's core values, once and for all, the Charter attracts those who will work to further New Urbanism's mission, but also insists upon and clarifies the terms of faithful allegiance and the boundaries outside of which New Urbanism cannot exist.

For those who merely work *with* New Urbanists and not necessarily within the movement itself, the Charter appeals to their best instincts. It works by piling minor agreement upon minor agreement until resistance collapses under the cumulative weight of things already acknowledged. Potential allies are persuaded by the details of a document whose virtues are attested by its adoption as the organizing principle for a unified movement.

The charrette as practice

"The charrette with its stunning efficiency and its democratic credentials is the one indispensable complement to the Charter of the New Urbanism." [Andres Duany, quoted on National Charrette Institute promotional materials.]

While the Charter clarifies the movement's position and solidifies the content of enrollment for members and allies, the charrette is the interactive tool that incorporates outsiders in the process of design and planning and works to ensure their ongoing support for the realization of the emerging proposal. The web site of the National Charrette Institute – an independent consultancy founded by early New Urbanists – defines the charrette as "a collaborative planning process that harnesses the talents and energies of all interested parties to create and support a feasible plan that represents transformative community change."[10] Practically, this entails three stages of work. First, pre-charrette publicity and planning makes everyone near the proposed location – from municipal officials to residents to local business – aware of the intention to plan a particular project and drafts possibilities for the site that are more or less responsive to future input depending on the designer. Second, the charette itself consists of intensive group discussions of particular, potential spatial arrangements of neighborhood elements. These are quickly rendered visually by facilitating design professionals in quick feedback loops. The discussions can take place in anything from a single week, to repeated two- or three-day engagements, monthly, for much of a year, depending on the inclination of the New Urbanist design firm leading the process.

Third, post-charrette work refines the plan and is intended to guarantee the continued support of the stakeholders.

The charrette is an acknowledgement of local, vernacular knowledge of how places do and do not work and a strategic approach to pre-empt resistance. Moreover, to borrow from David Brain (2005b), a sociologist and New Urbanist, the charrette de-specializes what Duany has called an over-specialized process: "the city is really not a city which has been assembled, it's ... a cartoon of a city ... the record of all the prerogatives of all the specialists" (Duany). Duany's only complaint about the charrette emphasizes its supposed virtue: in the excitement of getting to weigh in on general design issues, effort is required to make sure specialists address the issues in which they specialize.

The charrette enrolls allies for a project by acknowledging the importance to the process of anyone willing to get involved. Participants converge on a shared discourse by speaking up to design issues from a non-designer's perspective or speaking across design issues from the perspective of delimited design specialties. Lennertz, of the National Charrette Institute, referred to this as "the charrette creating the generalist approach from a group of specialists" (Lennertz). Note, this should not entail talking down, but communicating to the non-designer the concepts and vocabulary necessary to participate significantly. In addition, "anyone willing to get involved" ideally includes the anti-development NIMBYs who might interfere with the project down the line. The charrette should confront them and dissolve much of their resistance, both by respecting their input and by displaying the positive aspects of New Urbanist development. Lennertz said, "It's the delivery system," and goes on to assert that without valuing the charrette's core qualities, "collaboration and shared learning," a design process cannot be properly called New Urbanist (Lennertz).

Thadani, a charter member of the Congress for the New Urbanism, currently a member of its board of directors, and winner of the 2011 Seaside Prize, explained the enrollment, democratization, and de-specialization functions of the charrette in slightly different terms. First, he focused on trust and shared knowledge: "What we try to do is we build trust with the community, whoever the community is, whether it's the academic community [in the case of campus planning] or the citizens, planning department, whoever it is." Again this is not strategic duplicity, but a two-way process:

> what ends up happening is that we get a better understanding of the place. We start to understand things like where everyone likes to get their donuts in Lexington … start to really appreciate the local culture, which might be really different than what we're used to, but without trying, that just comes out. You talk about the greasy donuts, or whatever, in a presentation [of a final plan] and everyone knows exactly what you are talking about and you're not so much an outsider anymore … .
>
> Thadani

He claimed that the thoroughness of this trust-building process secured his former firm, Ayers Saint Gross, a "better hit rate."

Second, Thadani addressed cooperation and de-specialization, but as the deconstruction of the specialist habitus:

> The whole charrette process itself, if you do enough of them, really pulls back in terms of your ego, it's sort of a training. In a charrette, you give somebody a piece of a plan, you say, "Work on this." And then after they work on it for half a day or whatever you take it away and give it to somebody else and say, "Hey, you work on this." And they make some changes. And in a way what you're doing is *forcing history* – it's synthetic – because it's one hand, then somebody else is looking at it and modifying it – it's synthetic and in a way you're collapsing time by doing that. The person who you took it away from shouldn't feel – you sort of lose that attachment, that when you work on things you just can't help – you're just attached, that just the nature of it.
>
> Thadani

Here we see again the New Urbanist assumption of urban design benefits – non-Modern, organic synthesis, here – that motivate and accrue from de-specialization.

The charrette is at the core of the social process that is the New Urbanist built environment. As an interactive, collaborative design method, its primary characteristics have been taken up throughout the movement, unifying its members around an approach and generating a consistent effect where the movement meets the broader populace. The charrette constitutes a generalist design method that incorporates the interests and understandings of local laypeople, those who will be most affected by the construction of a project and also those most likely to resist it. By fostering generalism, it undoes decades of over-specialization that have contributed to dis-integrated sprawl – according to New Urbanists – to re-imagine profoundly *mixed* traditional urbanism. In both its commitment to democracy and its de-specialization, the charrette re-invents a vernacular.

Whether or not its champions' grand claims for its accomplishments are uniformly valid, the charrette has a clear role in the effort to consolidate a New Urbanist collective. In its central position as New Urbanist practice, it reproduces the field's commitments and organizes its habitus. Moreover, by incorporating laypeople, it regularizes, at least somewhat, the point of contact between the movement's members and a broader public who will profoundly affect its success or failure.

The Kentlands, Gaithersburg, Maryland

Recognizing how the New Urbanism is simultaneously a social and a material phenomenon requires that we turn now to the story of the Kentlands, in Gaithersburg, Maryland. Part of the Kentlands' significance is simple: the movement needed a model, something to show off to those it was trying to win over to its cause, and the Kentlands provided that. But another part of its significance is more complicated: in imagining, designing, planning, building, and selling the Kentlands, its creators found a concrete solution to the matrix of political, economic, cultural, and professional problems that shaped development in the suburbs. The specific physical details of the development itself resolved this matrix.

The concept of the Kentlands

The Kentlands is built on the former Kent Estate, a 352-acre property that, in 1987, was one of the last large, vacant parcels in Gaithersburg, and probably in that part of Montgomery County – an upscale, suburban area outside of Washington, DC. The Kentlands was the first New Urbanist, neighborhood-scale project built from the ground up in a conventional suburban context. The development was substantially complete by 2001, currently including approximately 1,800 homes, numerous parks, an elementary school, a mixed-use shopping district with restaurants, offices and a multiplex cinema, as well as a large, conventional shopping center at the edge. In 2011, the 340-acre, adjacent Lakelands – begun in 1998, and once part of the same property – was also substantially complete, including 1,410 residential units.[11] Kentlands and Lakelands share a neighborhood center and Lakelands is designed according to the same principles as Kentlands, by the same firm, but not to as demanding a standard of materials and traditional design intervention.[12]

The Kentlands was begun in 1988. This was about five years before the first meeting of the Congress for the New Urbanism, thus it was not originally conceived as "New Urbanist" in those terms. CNU co-founders Duany and Plater-Zyberk designed the Kentlands, and it has served as a point of reference for most of the New Urbanist projects that have succeeded it. Though Kentlands was pre-CNU, what Duany and Plater-Zyberk called "traditional neighborhood development" was not unknown at the time. Seaside, in particular, captured the attention of both the media and design professions for its commitment to walkability, mixed-use, and vernacular architecture.

The development of the Kentlands is a story made up of many stories. One story is the one its founders like to tell about inspiration, vision, and ambition. This is the story of a real estate developer who wanted to build something different, the architect and his team who helped realize a vision, a planner and an attorney who made the process possible, and the small town mayor who saw how to put his town on the map. But an equally important story is a more sociological one about how these individual, creative pursuits were constrained, enabled, and acted out within the existing political, economic, and cultural context. In this story, two professionals struggle together to fit a grand vision of community into the morass of local land-use regulation, and a developer seeks to keep his vision economically viable in the face of unforeseen changes in the financial environment. In the background, there is a story about the American dream of home, hometown, and home equity – about the people who wanted to live in this imagined community and their defense of their aspiration for it and their investment in it: deep culture meets governmental-procedural structure. And there are more stories and other ways of telling them.

Process: assembling a vision

There is some disagreement about where or how Joseph Alfandre, the Kentlands developer, and Andres Duany were brought together. Alfandre's attorney Joel "Jody" Kline described Alfandre's reading up on innovative planning and then setting off to find Duany in Florida, but everyone else with a memory of the occasion refers to an introduction of one to the other by David Wolfe, a marketer and specialist in homeowner's associations and "community management," with whom each had worked previously. It was early in 1988.

Everyone does agree, though, that Alfandre and Duany are together the source of the vision for the Kentlands. While the circumstances of their first meeting were somewhat inauspicious – a sort of blind date squeezed into the middle of a business trip for Duany – its significance was apparently immediately obvious. Bill Lennertz was one of DPZ's first employees and was traveling with Duany at the time. He emphasizes the sense of the meeting as the culmination of Alfandre's search for the right solution:

> ... the short of it was that Alfandre essentially said, "I didn't know what it was I was looking for, but I now know." He had a plan by another firm and he just wasn't feeling right about it for some reason and he didn't know what he was looking for, but, in my observation, this is not his words, it didn't really strike him in the heart. So it was like a conversion [meeting Duany]. From there on out, it was just arrangements.
>
> Lennertz

At that meeting, Duany apparently proposed a charrette and some dates. According to Lennertz, Alfandre replied, "I'm looking for something that creates a special place and what you're talking about is what I think this place could be."

This, however, was just the beginning. Conversations between Duany and Alfandre inspired a trip to England. Ellen Shoshkes, in *The Design Process* (1989), writes, "The contract stated that before the charrette was held, the developer would join DPZ on a visit to London to look at examples ... " (231). As it happened, they went *before* signing the contract, "more to test their chemistry than to look at models," according to Alfandre executive Steve Wilcox (ibid.). The trip itself was unquestionably significant. Duany's memory of it combines its exemplary and empathetic value, "We took Joe to London to really see great terrace houses ... we saw the Greenwich Hospital. He said, 'Why can't the mall [we are anticipating at Kentlands] look like this?' It was a great little trip. It *aligned* us" (Duany). In addition to this obviously social accomplishment, we should note the manner in which the "great little trip" also emphasizes the value of the real and the historical for the New Urbanism.

Beginning June 1, 1988, Duany and Plater-Zyberk led the seven-day Kentlands charrette. Duany says, "Our earlier charrettes [including Kentlands] were really spectacular ... circus-like. They involved more people and we were less proficient" (Duany). Lennertz describes Duany's bringing together his "dream team," including architects Dhiru Thadani, Patrick Pinnell, Alex Krieger, and civil engineer Richard Chellman, but also credits Alfandre's staff for both being involved and adapting to Alfandre's radical change in direction:

> Joe's team, Billy [Winburn] and Steve [Wilcox] and Sue [Whalen], the three musketeers, they were going, "Holy shit! We just spent all this money on a plan and Joe just trashed it and now we're going to do this?!" But they were great. They were a good team, but they were the ones, especially Steve, the financial guy, he was the one who had the most critical eye on it.
>
> Lennertz

In *The Design Process*, Shoshkes describes the Kentlands charrette day by day (1989: 232–50). While this level of detail is not necessary for our purposes, a few important themes should be drawn out of Shoskes's account. The charrette began with more touring, this time for the entire team, of nearby sites like Annapolis (Maryland), Georgetown, and other neighborhoods of Washington, DC – the regional architectural vernacular. This generated a shared experience of historical models to which the designers could refer in laying out Kentlands (232). That evening, Duany, very much the leader throughout, lectured on town planning, critiqued suburban sprawl, and explained DPZ's approach to design (233). While his leadership was essential, including frequent tie-breaking decisions, the collegiality and intensity of the process are emphasized by Shoskes, who enumerates the dinners together, alludes to the amount of espresso consumed, and records workdays ending at 1:30am.

Probably the single most important subject, recurring from day to day, according to Shoshkes's account of the charrette, is something referred to only in passing in most of my retrospective interviews: the mall originally planned for part of the Kentlands site. As soon as Alfandre bought the Kent property, he sold the proposed mall portion – approximately seventy acres, according to Duany, at the intersection of two local highways – for $17 million (Arkin 2004). Shoskes reports that Alfandre "entered into a joint-venture partnership with Indianapolis-based shopping-center builder Melvin Simon & Associates" (1989: 230–31), apparently a nationally-known retail developer. She also says, "Alfandre is not the majority partner in the mall, and [Simon's] approval is crucial" (Shoskes 1989: 235).

Alfandre served as a moderating influence on the designers throughout the mall negotiations, tempering their contrasting approaches and encouraging further engagement with the problem, so as not to alienate the mall developer

(Shoskes 1989: 239). A compromise was eventually reached after six days of drawing and negotiation: the mall "[would] be integrated with the town center. The town [would] be visible from the highway, with its own exit; there [would] be no parking lots between the mall and town; and the mall's townside façade [would] front directly on Kentland's [sic] Main Street and form a public square" (1989: 244–5). Moreover, the food court would be designed to serve as an interface between mall and town, drawing pedestrians in and shoppers out, while office buildings would screen the town's public space from the mall's parking lot (242). It would also fulfill all of Simon's original requirements.

The Kent property was divided into five smaller neighborhoods for the charrette. Each was subjected to DPZ's version of the charrette process, what Shoshkes refers to as a "round-robin … the same drawing is passed off from one designer to another" (1989: 234, see also Thadani on the charrette above). In this process, the designers composed neighborhoods from varying and regionally appropriate generic building types, responding to topography, and then amending their colleagues' visions in turn. Throughout the charrette, the designers would "walk the site for inspiration," especially after a difficult day.

The on-site charrette also included the opportunity to address the public's concerns. Designers met with local residents about alleged departures from city of Gaithersburg's previously developed Neighborhood Four master plan and presented the neighborhood composition to local builders. Access to the design process seems to have appeased the residents, while the builders' doubts about mixing housing types and price points were directly addressed by Duany: they asked "'Do people in single-family homes want to live across from town homes and quadraplexes?' Duany's answer is: 'It's been tested and it works'" (Shoskes 1989: 239). Moreover, because they were on site, the designers could always confirm details by stepping out the door: "'If we needed to know whether a particular tree was worth saving, we could send someone out to look at that tree.' [sic] says [Patrick] Pinnell" (Shoskes 1989: 235).

Although the charrette's primary focus was design, questions of cost and profit were never too far out of the picture. Shoskes describes the designers enthusiastically mixing house types on blocks once the street layout was established, but notes, "reality [was] never far away – Alfandre executive Steve Wilcox constantly calculate[d] the feasibility of various mixes of units, based on the number of units-per-acre of varying density" (1989: 241). On the second to last day of the charrette, "Alfandre's staff … put together a pro forma program [an analysis of the project's composition and use of land] that show[ed] the plan … still 500 units short and encroach[ing] on too many wetlands" (Shoskes 1989: 245). Wilcox put pressure on the developing vision to accommodate Alfandre's financial commitments.

The last day of the charrette ended with a public reception in tents on the front lawn of the former Kent mansion that included "Alfandre's symbolic presentation of the Kent estate buildings to the town" as the home for a new arts center and function space as well as a presentation by Duany in which he showed many renderings of Kentlands as imagined by its designers (Shoskes 1989: 250). Jennifer Russel, then director of planning for Gaithersburg, remembers, "Andres presented this wonderful slide show to show what Kentlands was going to be. People were totally overwhelmed, 'Oh my god, it's going to be fantastic!'" (Interview with the author June 21, 2004 [hereafter "Russel"]). Bill Lennertz suggests that this final event was important for cementing support for the project, "Joe really did it up, had a big party out on the lawn in front of the barn. ... He leveraged it" (Lennertz).

The process of designing and planning of the Kentlands was a process of assembling a shared vision, then anchoring that vision in specific details. This is a complex process that involves a variety of methods and modes of interaction and a commitment to design. Perhaps foremost among the methods for assembling the vision was the use of built examples to produces a mutual understanding of issues of form. As Duany explains of a trip like the one he took to England with Alfandre, "It aligns the developer and the designer to discuss something with the physical example present. You can talk yourself into thinking you agree, but its only when you are confronting *things* that you decide, 'Oh is that what you meant ... " (Duany). Tours of historical models like Annapolis and Georgetown early in the charrette played the same role for the charrette team. Moreover, this value placed on built examples implicitly recognizes how the new place will itself function as an argument for the principles it embodies – if the designers could learn from existing cases, so can their successors.

Then the on-site charette transformed a general agreement into a specific plan through the work of skilled professionals who themselves had to come to some consensus. They confronted not only the contrasting ideas of their colleagues, but the constraints of working for a for-profit developer in a rigid regulatory context. This meant condescending to incorporate a mall into a traditional small town, wrestling with parking and traffic issues, and controlling costs. It also meant avoiding certain regulatory traps, such appearing to divide and develop wetlands, which would require the involvement of the Army Corps of Engineers and an additional layer of review and permitting. As various of my interviewees have said, the greatest benefit of producing the plan in a group this way is that it creates many champions for a vision that will need enduring support to be realized. Again, it does that by capturing consensus or compromise in specific detail – how the mall will relate to the adjoining neighborhood, how many houses and of what assorted sizes will be built on a particular block, how the neighborhoods will work around wetlands. As we will see, it can be essential that these champions come from the public as well as the ranks of the professionals involved, and thus are the locals welcomed into the process.

Product: the Kentlands as model

We will return later to the story of the Kentlands' completion. Once underway, however, Kentlands became integral to the ability of the New Urbanism to reproduce itself by serving as proof of its capability. The New Urbanism was not just a bunch of ideas, but a plan a developer could complete, a fact on the ground. When asked about the relevance of the Kentlands for the constitution of the movement in its earliest stages, Duany said that it was not marked, because the

Kentlands was not far enough along in the early 1990s and that only Seaside was sufficiently complete to serve as an inspiration. But, when asked about the relevance of Kentlands as a model for developers, for the ongoing reproduction of the movement in its middle stages, Duany acknowledges Kentlands' importance:

> That has to do not so much with design, although it's not that bad. It has to do with proximity to a good airport and completion. We have lots of projects in the pipeline, but they're ten years out. And also … [Kentlands] does all three [types of retail]. It does the main street. It does the ordinary retail of the Square [shopping center]. It does big box … . Pretty interesting. Not particularly gracefully designed together, sort of cobbled together when they bought Lakelands, but pretty interesting. Lot of lessons there.
>
> Duany

Duany says he frequently takes developer clients there.

Russel suggests that the Kentlands, once built, has been essential in conveying the central tenets of the New Urbanism to developers, planners, and municipal officials. It has become a place to go see. Because of her involvement in the project, firsthand knowledge, and pioneering experience, from a planner's perspective, Russel describes how she became the person to call about New Urbanist planning.

> Oh yeah, oh yeah, constantly … we had this whole – well first of all, each time the APA [American Planning Association annual meeting] was in Washington, we've had all kinds of big tour groups. In the spring … Mike Watkins [of DPZ and the Kentlands town architect] and I used to think we were like travel agents. We did a big job for certain groups, not everybody, but elected officials, planners from all over the country. I would sometimes try to get the mayor involved or the city manager involved. We'd take them on tours, we did packages, I don't know if they're doing that now, because that was like a big thing when I was involved. … Clark [Wagner, a member of the planning staff] and I used to do all the tours. Well, he left a couple of years ago, and then I tried to start training people to do the tours because sometimes it got so I couldn't do all the tours. Did a lot of tours, a lot of requests for information, which we gladly send people.
>
> Russel

Russel went on to explain that visiting city planners were always interested in all the innovations, while developers were enthusiastic about the feel of the place, but concerned about whether they could manage the infrastructure costs. Elected officials, she says, were split over the issue of density – some loved the Kentlands and others could not believe anyone would want to live so close to anyone else.

Obviously, the Kentlands has had a profound effect as a model for those directly involved in planning, permitting, and building new places. Because it stands as a concrete example, Kentlands is a formidable tool for the continuous and repeated recruitment process. Moreover, like Alfandre's trip to England with

Duany, it seems that experiencing the place helps to persuade people about the principles of New Urbanist design and planning.

Conclusion

Chapter 2 established the pertinent political, economic, cultural, and professional conditions of the post-postwar period within which the New Urbanist movement would emerge. This chapter has investigated the processes internal to the constitution of the movement, highlighting social capital and relationships, discursive mechanisms and practices, and integral place-making events. The founders' narrative emphasizes the significance of elite educational institutions to this story, but balances that with a sense of interpersonal affiliation and friction. The discussion of the Charter shows how discursive positions can be claimed, consolidated, and reproduced, while the charrette becomes a practice that defines a field, in straightforwardly Bourdieusian terms. The account of the Kentlands' origins reminds us of the importance of interpersonal and inter-professional dynamics, but also locates them within the process of making a place, anchoring interactively negotiated solutions in concrete details of the built environment.

This deeper, more detailed sense of the movement and its process of becoming enable us to pan back out to the larger scale, locating the manner in which the New Urbanism responded to the historical conditions explored in the previous chapter. We will see that these micro-processes all occur within a complex and interconnected environment that constrains aspects of the movement, while facilitating others. Moreover, these micro-processes, while influenced, are simultaneously integral to the reproduction of the macro-processes that shape them.

Figure 3.1 Single family houses on Lake Street in the Kentlands

Figure 3.2a Pedestrian paths in the Hill District, the Kentlands

Figure 3.2b Pedestrian paths in the Hill District, the Kentlands

Figure 3.3 Row houses and a small playground, Hill District, the Kentlands

Figure 3.4 A Kentlands Victorian and parallel parking

Figure 3.5a Row houses in the Kentlands

Figure 3.5b Row houses in the Kentlands

Figure 3.6 Row houses and apartment building in the Kentlands – note: this is a rare mix of housing types in American suburbia

Figure 3.7 Apartment building in the Kentlands

Figure 3.8 Single-family houses on Booth Street in the Kentlands

Figure 3.9 The Main Street Pavilion in the Kentlands

Figure 3.10 Mixed use on Main Street – apartments above shops

Figure 3.11 More mixed use on Main Street

Figure 3.12 Density just off Main Street in the Kentlands

Figure 3.13 A new office building on Main Street between the Kentlands and Lakelands

Figure 3.14 Two-over-twos in the Lakelands – a near-unique housing type: two two-floor units, stacked vertically

Figure 3.15 Row houses in the Lakelands – note the lower quality in the asphalt shingle and design elements

Figure 3.16 House facades in the Lakelands

Figure 3.17 Single-family homes in the Lakelands

Figure 3.18 Single-family homes and large row houses in the Lakelands

4 Reimagining the suburbs, beginning to build

The post-World War Two production of suburban built form entered a second phase in the 1980s. The original postwar phase was effectively defined by the response of the "community builders" – large-scale, residential developers (Weiss 1987) – to the postwar housing shortage, the availability of federal mortgage insurance, and the White, working- and middle-class exodus from American cities (Fishman 1987; Hall 1988; Hayden 2003; Jackson 1985). The second – newer, post-postwar – phase exaggerated and consolidated the tendencies of the first by building bigger single-family homes on bigger lots in strictly residential pods in further-flung and increasingly automobile-dependent suburbs.[1] These post-postwar expansions of the original elements of the suburban boom were complemented by the innovation of "category-killing" big-box stores, retailers that overwhelmed traditional department stores by providing extensive selection in a few categories at lower prices. These trends were futher reinforced by the introduction of novel financial instruments to fund real estate development, such as the growth of a secondary market in mortgage-backed securities (Leinberger 2005, 2000); the further withdrawal of city planning from control of form in favor of regulating use (Blau, La Gory, and Pipkin 1983; Fishman 2000; Hoch 1994); and the active self-absenting of architecture from the suburban development process (Blau 1984; Cuff 1991; Gutman 1988; Larson 1993).

Chapter 4 explains the interaction between the "what" and the "when" in the driving question "Why New Urbanism now?" It analyzes how members of the New Urbanist movement engaged with the cultural, professional, and financial context of the 1980s and 1990s as a set of opportunities and constraints in relation to their efforts to establish and expand the movement. It does so from an interior perspective, from a point of view that understands the movement as an intentional actor, revealing how the movement's leaders understood their interventions. By locating the emergence of a novel approach to producing the built environment in its social-historic context, this story sheds light on how specific spaces emerge in interaction with the dynamics of their moment. The story of "Why New Urbanism now?" is a story about the active and interactive relationship between the political, economic, cultural, and professional context and a collective effort to alter that context. Or put slightly differently, how did the emergence of the New Urbanism described in Chapter 3 work within the constraints and opportunities inherent in the moment to emerge as one answer to the question of how to do suburban development differently?

Recruiting the professions: architects and planners

New Urbanists capitalized upon frustrations within the fields of planning and architecture by offering a route back into meaningful roles in the process of suburban development. In the period between World War Two and the early 1980s, planning had its degree of authority, its area of influence, and its professional autonomy radically undermined by changes within the field and the disastrous failures of urban renewal policies with which the profession was complicit. The New Urbanism offered planning a renewed control of form and the chance to plan *good* places, rather than simply managing potential conflicts of use. Simultaneously, it invited architects to forgo some of the creative freedom the profession privileged in exchange for more influence over the places in which most Americans actually live. Critically, the New Urbanism did these things through the specific details of producing places.

Architects and the charrette

The process of building the New Urbanism began with architects: groupings of architects with a variety of complaints and interlocking social and intellectual connections, interactions which gradually turned into a decade-long, charrette-like search for a vision of alternative ways to approach development. But the architects at the core of the New Urbanist movement confronted a tremendous obstacle in their need to recruit their professional fellows to expand the movement. The fundamental principles of the New Urbanism contradicted the definition of the modern architecture profession, particularly its commitment to unbridled, artistic freedom and the single, sculptural building (Larson 1993).

Strategically, New Urbanists capitalized on the absence of architecture from the conventional suburban development process by drawing on a practice that they still had in common: the charrette. The charrette became, at various scales and in various applications, the technique by which the movement incorporated architects disaffected by the frustrations of modern architectural practice and the model for the New Urbanism's broader dissemination. This expansion of the charrette concept was foreshadowed by the "Pedestrian Pocket Charrette" discussed in Chapter 3.

The New Urbanist charrette could accomplish this enrollment of architects because it pursued and facilitated development. New Urbanist principles might sound good to architecture students frustrated by theory, Modernist orthodoxy, or Post-Modernist incoherence, but the proof would be in the pudding. New Urbanists could attract architects if they could build. And build they did: the Kentlands project became a demonstration that this process could move from a vision to actualization, from charrette to – eventually – place, even if that place was, by some lights, stylistically banal and bourgeois. The Kentlands charrette cut the Gordian knot of land-use zoning, housing finance, and NIMBYism by bringing all the stakeholders into the same process. As a concrete basis for New Urbanist principle, moreover, the Kentlands grounded their advocacy efforts and gave them a point of reference from which to adapt to various situations.

As Latour said, " ... the easiest means to enroll people in the construction of facts is to let oneself be enrolled by them!" (1987: 110). The New Urbanism's founding architects expanded the movement by making themselves and their principles available to be enrolled by disgruntled architects looking for a new direction. That enrollment, furthermore, began a process requiring of all members a change in *habitus*. As Thadani described in Chapter 3, the charrette demands deconstructing the specialization of everyone involved. "The whole charrette process itself, if you do enough of them, really pulls back in terms of your ego ... ," he pointed out (Interview with author, November 11, 2005). Participating in the charrette, they were sacrificing their mastery of the design process; participating in the larger process that was creating the New Urbanism, they were joining a collective that redefined the role of architect. It was not that design skills were no longer relevant, but that they were one of a set of skills that New Urbanist practice required.

Becoming a New Urbanist architect can provide for some a resolution of the tension between the primacy of style in the field and the question of usability. It expands the scale of the design beyond the limiting ideal of the iconic building, returning to a broader focus on the metropolis as a whole. This broader focus for urban design, on usability and community as goals, increased architecture's relevance and addressed the desire among some architects for a progressive social purpose. By exchanging the nominally free exercise of creativity for the primacy of historical precedent in urban form and vernacular design, architects working within a New Urbanist framework gained a claim to normativity – to be designing good places, because sense of community is widely seen as a virtue. Finally, New Urbanist architectural practice is a way of "reskilling" the experience of working as an architect (Gutman 1988), primarily through the charrette, because early-career architects gain the opportunity to contribute substantively to design in the collaborative process, rather than functioning merely as draftspeople.

Murphy Antoine is a principal in a multiple award-winning, New Urbanist design firm, Torti Gallas in Silver Spring, Maryland. He is representative of my interviewees among New Urbanist architects in his experience of architectural education as overly concerned with style and aesthetics and comparatively inattentive to social function and responsibility. Furthermore, Antoine's account of his early career architectural practice, both fits with Gutman's sociological analysis of the profession and captures the economic realities of architecture as a service profession. He explained:

> If school was all about style and [working for a production builder] was all about the economics and the market forces of suburban development, then why do I even want to do [architecture]. ... [But] New Urbanism tied all of that up for me."
>
> Interview with author, November 2005

The disillusionment of some architects with the state of their profession provided an opening for the New Urbanism to gain acceptance that otherwise would have been closed to them because of their rejection of some of architecture's fundamental commitments. This disillusionment grew from the disjuncture between an architectural education that elevated creativity above all else and – early career – architectural practice that emphasized service and craft. New Urbanist architects turn away from the ideal of unfettered artistry in favor of a respect for precedent to gain a normative position and a degree of relevance, both in their own early-career practice and for the majority of the homebuying public.

The charrette thus turns out to be not only the organizing process which led to the creation of New Urbanism and its primary method for creating the enrollment necessary to generate and realize visions of specific developments, but also an emblem of the professional transformation architects permitted and required of themselves to that end. The early congresses established this interactive approach as core to the movement. The Charter established the movement's commitments, but also broadened the scope of the meetings, transforming them into more conventional professional meetings, less the de-specializing exchange across disciplines in the pursuit of successful design. The combination of charrette and Charter effectively constituted a field, in Bourdieu's sense, with a field specific habitus, to which the charrette is central.

Planning's crisis of confidence

The planning demands associated with the postwar suburban boom hastened a shift from practical experience with the control of urban form among working planners toward a policy-oriented, numerical-rational, educational formation (Fishman 2000; Hoch 1994). Educational programs were developed, in part, in response to shifts in theoretical discourses in planning, but, additionally, with "demographic" pressures in mind – the need to get trained planners out into the world (thus undermining the value of practical experience in planning) (ibid.). In relation to this abstract turn, the profession caught architecture's fascination with the abstract clarity of Modernism in a way that shaped emerging, postwar efforts at slum clearance into "urban renewal." When authoritarian urban renewal's negative effects drew public ire in the 1960s and 1970s, a planning establishment recently turned toward abstraction was forced to cede much of its remaining control over form to the due process of citizen board oversight and public participation. From inside the profession, Fishman and Hoch both show that planning, as a result, was radically diminished in its ability to affect the production of the built environment, even in its own terms (Fishman 2000; Hoch 1994).

Nevertheless, planners retained a gate-keeping role within the development process, controlling access to permitting and approval throughout the United States. Anyone who wanted to take on the large-scale design of urban space would have to convince planners of the merit of their ideas. In addition, the so-called Euclidean zoning that planners have conventionally implemented continues to make many elements of New Urbanist designs illegal, thus drawing out the permitting process and increasing its expense. Thus the architects at the core of the New Urbanist movement had to recruit at least some of their planning colleagues to progress in the realization of their goals.

The effort to persuade and involve planners required a balanced approach – another constraining opportunity. On the one hand, the New Urbanism had to provide a way for planners to accomplish goals whose unattainability had frustrated them. Opening the New Urbanism to planning meant demonstrating to planners that, even though it ran counter to the majority of their current practices, the New Urbanist method had merit in planning's terms and could help resolve some of the profession's perennial trials – that it improved upon current practices. On the other hand, the New Urbanist approach had to avoid challenging the basis of planning's control over the development process – the implementation of a code. New Urbanists had to acknowledge planners' expertise in matters regarding the implementation of complex regulations. Finally, while appealing to planning, the New Urbanist approach would also have to be politically viable in the broader context of local politics that governs municipal planning.

The New Urbanist response to this need to enroll planners came together in the "form-based code." Growing out of the prescriptive, graphic codes of the earliest New Urbanist projects like Seaside, form-based codes, as their name suggests, shifted the regulated quality of the built environment from use to form and shifted the emphasis from prevention to encouragement. This meant conceiving first of how planners wanted a space to perform – as a busy, transportation-accessible, downtown, employment and entertainment district or as a quiet residential area or as a small, walkable, village commercial center – then specifying the form of the space to promote those dynamics – using street and sidewalk width, building height and type, etc. – and leaving the uses to the occupants.

The point of form-based codes is that regulators' priority should be to integrate uses in positively useable spaces. This shift in emphasis in planning has been made possible by the advent of a substantially post-industrial era for American metropolises, where containing the hazards of industrialization is no longer the first priority – the original motivator for land-use zoning. According to New Urbanists, use-based zoning defines a paradigm of avoidance that has outlived its purpose and has become both inflexible and inaccessible in its Byzantine accretions.

This account of the status quo and the profession's potential – and form-based coding – have proven appealing to planners. The late Paul Crawford, planner, form-based code expert, and co-chair of the CNU's Planners Taskforce, remembered his initial interest in the New Urbanism in terms of this effort to replace an obsolete approach: the conventional "zoning code did so little to shape the nature of development appropriately," while "New Urbanists were trying to write codes to produce the *desired quality* of development – quality *and* character" (Interview with author, June 10, 2005).

Form-based codes popularized by New Urbanists have not asked planners to give up their tools, but rather to take up revamped ones better suited to a new understanding of their professional purpose – the crafting of good places, not the avoidance of conflicting uses. Peter Katz, original CNU director and author of *The New Urbanism* (1994), remembers:

> What's so interesting about the codes, as long as the New Urbanism was a debate, planners were not terribly interested, because planners get enough

> conflict in their work. But the New Urbanism suddenly moved from being a debate to becoming a technique for implementation that was easier, more efficient. Then they were all over this ... it's a whole new game ... .
>
> Interview with author, November 29, 2005

These codes preserve the basis of planners' control over the process, but reframe the method of that control and its tone. Moreover, they can improve planners' relationship with the citizen bodies that oversee their work.

The process of creating a form-based code reflects and incorporates the methodology of the New Urbanist design charrette. As we have seen, the charrette as public process can include local stakeholders, draw on local knowledge, and incorporate public feedback. Ideally, this can identify and confront sources of local resistance to the project early and build a constituency interested in its implementation. Lastly, the charrette does need a leader, and planners, as local officials and professionals, are well situated to fulfill that role. It thus maintains planners' control of the process.

In sum, form-based codes as an element of the New Urbanist approach offered planners the opportunity to take up issues of physical design that they had professionally forsaken, manage the public process, and reassert their professional distinction without giving up their gate-keeping roles as the implementers of codes. In this sense, form-based codes "lashed-up" a range of interests and prior commitments in a specific – nearly – material articulation. This can be seen as the pragmatic seizing of an opportunity by the New Urbanism. Given the entrenchment of the planning process, the predominance of Euclidean zoning, the planning profession's location at the center of the zoning process, and the profession's complicated recovery from its crisis of confidence, any successful approach to change would need to alter planners' goals without interfering with their role. Doing otherwise would have constituted a direct attack on the profession's expertise and remaining authority.

Zoning constraints also, then, allowed the architects who started New Urbanism to create a situation in which the planners, boxed into their narrow role, could seize the opportunity to find a new outlet for pinched creativity. Long used to dealing with local authorities, more recently forced to deal with the public, they had skills they could bring to bear in running the newly reconceived charrettes that architects lacked. Once enrolled, planners could take over the coordination problem, leaving architects freer to concentrate on design. Here, too, the charrette proved central to the process of enrollment and to the eventual development of a new habitus for planners.

Reclaiming urbanism: urbanity, sprawl, sustainability, walking, and community

How did New Urbanists reframe the discussion of development so as to enroll all the other players needed to realize their vision on its various levels? They needed to engage with the portrayal of the suburbs – their advantages and their shortcomings – but to propose concrete solutions to the puzzles they confronted.

Reviving "urban" and framing sprawl

The New Urbanism's expansion has, in part, depended on its ability to unravel a key cultural tangle – the suburb as node of cultural conflict. By redefining urban, without denigrating suburban life, New Urbanists have managed to criticize suburbia without suffering the popular cultural backlash that elites have traditionally faced when expressing similar criticism. The New Urbanists have accomplished this by shifting the terms of the debate to focus on "community," a shift that both avoids the usual, problematic mass–elite opposition, and turns the focus instead to design issues. A focus on urban form has also enabled them to respond to suburban frustrations with auto dependence and various environmental concerns.

Importantly, these New Urbanist interventions are not discursive countermeasures, but designs, addressing predominant discontents by improving upon the suburban spatial status quo. The point here is that the movement has accomplished a sort of cultural-political judo grounded in the specific details of the built environment: they have turned widely shared dissatisfactions into motive forces, driving New Urbanists and the New Urbanism forward and into a broader public consciousness.

The devaluation of the character or conception of urbanity was implicit in our earlier discussion of the postwar suburban shift. "Urban" had a dingy, pre-World War Two, working-class, ethnic resonance, precisely what the postwar exodus was escaping. Those residents who left pursued instead middle-class status and unambiguous whiteness. This anti-urbanity underpinned many of the phenomena previously discussed including the emphasis on the single-family house, the segregation of uses, and the wholesale destruction of neighborhoods that characterized urban renewal. Moreover, the commitment to the not-urban or suburban – particularly in the new, single-family home bias built into mortgage assistance – reproduced and reinforced the degradation of the real city, effectively guaranteeing postwar disinvestment. Nevertheless, as the limitations of the conventional suburban model became more conspicuous, from the 1970s into the 1980s, the possibility opened up for the reconsideration of the urban.

One of the first things that the New Urbanism did, apparently quite consciously, was to reclaim the word "urban" and positive associations with urban environments, through its appellation, of course, but also through its presentation. In my interview with him, Ray Gindroz, co-founder and principal emeritus of Urban Design Associates and former CNU board member, argued that, before the first CNU congress in 1992, "'urban' was still a four letter word." By appending the word "new" and pairing the concept with "images of sparkly, clean things" – i.e., with renderings and photos of the earliest neotraditional developments – the movement began to undo the concept's negative connotations (Ray Gindroz, interview with author, June 9, 2005). It was essential, moreover, that this reclamation coincided with and capitalized upon a resurgence in the livability of important American cities like New York, while supplying an alternative to those same cities.

The reclamation of "urban" also drew upon the widening and deepening critique of conventional suburban development, discussed in Chapter 2 and most succinctly captured in the popularization of the epithet "sprawl." The use of "sprawl," and the less pejorative "conventional suburban development," is a consciousness-raising rhetorical device for New Urbanism and a number of allied anti-sprawl movements – especially Smart Growth, environmentalism, and farm preservationism. Sprawl is inherently negative, but even noting the conventionality of conventional suburban development – that the familiar suburban environment requires a qualifier – contains the assertion that it is just one alternative, perhaps not the best one.

Thus, in calling themselves New Urbanist, the movement implied that while urban need not refer only to the decaying city centers of the 1970s, suburban should not refer exclusively or uncritically to the semi-pastoral American ideal or its recent MacMansion manifestation. In reclaiming urban, the New Urbanists were claiming that there could be an alternative to what had come to seem the only appealing option for most Americans. The New Urbanism asserted that the public might choose differently if it were presented with a choice. Elizabeth Moule describes how her interactions with suburban mothers confirmed her sense of possibility for the movement:

> [M]ost of them, and these are people with significant means, find themselves quite trapped in this life that they have elected to make for themselves. A lot of people have a hard time imagining that you could do some things and it could change the complexion of the built world ... most people felt resigned to whatever it was. [But] As we were talking with people and making them understand the connections from A to B, it was suddenly quite obvious to any halfway thinking person what the situation is ... in city after city after city, on project after project after project, what we were finding was there was an enormous grassroots support for what we were talking about.
>
> Interview with author, January 26, 2006

Because the suburban was coming to seem less than ideal among upper-middle class homebuyers, the urban could be brought closer to parity, elevated, by New Urbanists, but carefully. Not coincidentally, the most frequent concrete manifestation of the New Urbanism is in suburban greenfield developments that purport to borrow the best of urban spatial dynamics without asking a still-hesitant population to actually move to the city.

The promise of sustainability

The New Urbanism also found an opportunity in the environmentalist resistance to development established in the wake of the postwar period. While some New Urbanist design firms, particularly that of founders Moule and Polyzoides, specialize in environmentally sustainable building techniques, the basic environmental argument for the New Urbanism generally derives from the reduced physical footprint of New Urbanist development and the reduced environmental footprint of its inhabitants. Dense development requires less land, leaving more open and undeveloped. There is also less paved area in New

Urbanist developments – with their narrower streets and smaller parking lots – thus reducing impermeable surfaces and easing storm-water management issues and reducing the heat load. The greater frequency of greenspaces and the use of street trees also improve the environmental performance of New Urbanist places. New Urbanists have promoted their "green-ness." They argue that neighborhood design can reduce the carbon footprint of residents by decreasing the number of car trips necessary to fulfill daily needs. The New Urbanism's recent alliance with the United States Green Building Council (USGBC) is a kind of stamp of approval. The USGBC is the creator of the influential Leadership in Energy and Environmental Design (LEED).[2]

The claim that New Urbanist development is more environmentally responsible than conventional development aids its progress in a number of ways. First, it attracts some element of the homebuying market concerned with sustainability. Perhaps more importantly, like the trope of community, the promise of environmental sensitivity can also diminish local opposition to development by directly addressing one major basis of resistance. But unlike the vague promise of community, the claim of sustainability must be carefully demonstrated, thus the partnership with an already well-known rating organization.

Despite these efforts, the movement still frequently confronts and conflicts with environmental regulations put in place in the response to the suburban boom. Until very recently, New Urbanist developers have been required to re-zone the land they plan to build upon because New Urban designs conflict with conventional zoning codes, a time-intensive process. On top of that, Environmental Impact Reports, conceived to prevent the thoughtless clearing of natural habitats associated with conventional suburban development, slow the process still further. Similarly, wetlands preservation rules, again intended to focus attention on use-values being lost for other species and the effect of development on ecosystems, break up the connectivity upon which New Urbanist plans depend to achieve walkability and disperse traffic. Duany explains:

> So they [environmentalists] have to understand that the way to save nature is to make cities great and that they have to forge new tools that aren't just the green ones. It isn't just about water quality. It isn't just about more trees. It isn't just about bikeways. *It's actually about making cities like Manhattan.* And environmentalists need to become experts in making such places because, at the moment, *their well intentioned laws inadvertently*, and this is what's curious, *prevent such places as Manhattan from being built.* Because the maximum lot-coverage rules [related to storm-water management] and the 'leave the streams alone' rules clip the grid [of streets necessary for New Urbanist connectivity].
>
> Interview with author, November 8, 2005

The only recourse, according to Duany, is to persuade environmentalists to get behind the New Urbanism: " … there's no power even remotely as powerful as the environmental groups, it's just going to be the driver of the future, period," and in the end, the New Urbanism will "prevail because the environmental movement's going to get behind it" (ibid.) Duany explained to me that this was his next project, getting the environmentalists on board, and it seems to be occurring, particularly through LEED-ND.

Get out and walk

Suburban Americans' frustration with the interrelated experiences of traffic congestion and the necessity of frequent car trips has formed a hugely important opportunity for the New Urbanism. New Urbanists argue that conventional suburban form – its segregated uses in particular – and street network and highway design mandate not only car trips for every purpose and worsen traffic, but actually prevent walking. Conventional suburban development and street design make walking long, boring, unpleasant, and dangerous.

Various studies in the planning literature debate whether or not people living in New Urbanist developments or the traditional neighborhoods they emulate, as opposed to conventional suburban ones, actually walk more and drive less (Crane and Crepeau 1998; Handy 2005; Handy and Mokhtarian 2005; Lund 2003; Song and Knapp 2004; Southworth 1997),[3] but for the purpose of building the movement, the observed behavior is less significant than the movement's appeal to a vision of a different and better life. People claim to want to be able to get out of their cars and New Urbanists claim to offer them that possibility. At the same time, suburban residents do not want to be forced to give up their cars, so they need to continue to be accommodated.

According to New Urbanists, pedestrians should experience the street as a well-proportioned "outdoor room," an effect achieved through managing street width, building height and articulation, and the proximity of buildings to the street. Pedestrians should, moreover, encounter a varied streetscape – with differing, but consonant, architectural details from building to building – and visual goals for routes – vistas, for example, should be terminated so progress is perceptible and should focus on notable buildings. These elements are missing in conventional suburbia, according to New Urbanists.

To make walking appealing also means making it seem safe. New Urbanists – drawing on Jane Jacobs's concept of "eyes on the street" (Jacobs 1961) – assert that people need to feel that they can be seen as they walk, to feel safe from crime. This is simple to achieve, according to New Urbanists: bring buildings close to the sidewalk and street and require adequate fenestration, including the transparency of windows into shops or restaurants in commercial areas.[4] But a sense of safety also means protection from cars. Narrow streets, parallel parking, street trees, and alley-loaded garages – as opposed to curb-cutting driveways – can calm traffic and keep cars a safe distance from pedestrians. All of these elements of "walkability" occur in historical places, and New Urbanists want to reproduce them because of the benefits they claim to have observed.

In addition to walkability and the obvious environmental benefits of driving less, the New Urbanists make an economic argument for the reduction of car use. In *Suburban Nation* (2000), Duany, Plater-Zyberk, and Speck cite an American Automobile Association estimate that the cost of owning a cheap car is more than

$6,000 per year on average (56). They estimate that by eliminating the necessity for one car, a household could afford approximately $60,000 more house (ibid.). In an article in *Urban Land*, the magazine of the developer's professional organization, the Urban Land Institute (ULI), Christopher Leinberger – a New Urbanist finance guru – estimates the potential savings could yield twice that, more than $120,000 in mortgage loans (Leinberger 2000). In this way, the New Urbanists have found another angle – one particularly appealing to the average homebuyer – from which to argue the reduction of car ownership. The "location efficient mortgage" is a home financing product developed in response to this argument, making greater funds available to buyers in locations that will minimize the necessary expenses of private transportation. In fact, by 2006, the location efficient mortgage was only available in Seattle, San Francisco, Los Angeles, and Chicago.[5] Relating car use and affordability of homeownership also takes on a new importance in an era of unstable gas prices and record foreclosures.

Actually accomplishing walkability in new suburban development is difficult for a number of reasons. For example, to foster variety in a subdivision, New Urbanist designers must ask developers not to sell large blocks of adjacent lots to a single builder, but rather to group lots back to back so that when builders produce similar houses, they will occur on different streets, not next to one another. This increases construction costs by mandating multiple delivery sites for materials and other measures that run against assembly-line-like efficiency. In fact, it makes some bigger builders less willing to work in New Urbanist developments.

Similarly, New Urbanist designs proposed to increase pedestrian safety must pass the review of municipal officials responsible for traffic, emergency services, snow removal, and trash collection who are used to streets designed to perform at higher speeds. Dhiru Thadani says:

> The traffic engineer goes hand in hand with the fire chief who wants wider streets, bigger toys. The trash collection people – you can see this everywhere, the same trucks go through a place like Alexandria with small roads, but their preference is not to do that because they have to go slower. So if the road is wider and clearer, yes, it makes their work easier.
>
> Interview with author, November 11, 2005

For New Urbanists, passing this review depends on demonstrating, among other things, that narrower streets and smaller corner radii can still accommodate the service vehicles that require neighborhood access – and that smaller, usually European, service vehicles are available (and in use in Charleston, South Carolina, according to Thadani) (ibid.). My informants consistently report that the objections of public works and fire officials remain their greatest hurdle.

Finally, to produce the walkable places they promise, New Urbanists must also negotiate conventional zoning codes that disfavor mixed-use and density so that people will have somewhere to walk to. Integrating the commercial, residential, and recreational functions of a neighborhood is necessary in order to bring worthwhile destinations within walking distance.

Density, it turns out, is also necessary: commercial functions, schools, and, ideally, public transportation require significant populations close by to sustain them. This is especially true since what walkable means to most Americans, according to New Urbanists and others, is distances of one quarter to half of a mile, approximately a five to ten minute walk. But density means convincing American homebuyers that they need less space than they have come to expect. New Urbanists attempt to do so by substantially eliminating front and side yards, maintaining backyards, but not decreasing the size of homes – average interior square footage is nearly the same as in nearby conventional developments according to Tu and Eppli (Tu and Eppli 2001) and the reports of realtors active in the Kentlands.

The trope of community

The New Urbanist response to the range of suburban discontents and criticisms discussed above comes together in concrete designs for new places laid out like pre-World War Two small towns. The concrete and discursive come together in this reference: a resemblance to pre-war, small-town America has as much to do with our emotional associations as with physical form. Whether traditional small towns encouraged greater senses of community than contemporary suburban subdivisions is a matter of some debate (Putnam 2000; Skocpol 2003; Skocpol and Fiorina 1999), but the popular image of small towns certainly includes it. The New Urbanism thus plays on conventional suburbia's soulless reputation, without taking up urban elites' disparagement of contemporary suburbia: they transcend the suburban culture debate by claiming to produce "community" through design.

The added benefit of relying on such a nebulous concept as community is that neither the New Urbanists themselves, nor their customers need to be able to define community – which also makes proving its absence more difficult. New Urbanist architect, academic, and internal critic, Ellen Dunham Jones goes a step further:

> The New Urbanists use the term community building without really clarifying. Are they talking about the physical community or are they talking about social community? They will tend to often deliberately allow for or include that ambiguity, because they want to believe it's one and the same.
>
> Interview with author, November 2, 2005

The claim to produce a sense of community may work, moreover, to attract the support of those beyond the homebuying public. It can decrease local resistance to development and draw the interest of professionals. Community can redeem new development emotionally or psychically for the neighboring public by promising to introduce a public good. It is harder for NIMBYs to foster the outrage necessary to interfere with the building of a project when it is presented as a good place. In addition to appealing to local laypeople, Yale architecture professor Allan Plattus suggests that, early on, the claim to produce community attracted elements of the design profession: "that it could help, directly and instrumentally, rebuild community – wow! What could be better?" (Interview with author, November 15, 2005).

Two factors appear relevant to the New Urbanists' success in producing a sense of community in their developments (cf. Brown and Cropper 2001; Harvey 1997; Kim 2001; Langdon 1994; Rood 2003; Talen 1999) – though whether community is itself a desirable goal is also a matter of some debate. The first is self-selection: early New Urbanist neighborhood-scale developments constituted such a small proportion of suburban real estate, were so distinct from conventional development, and commanded so obvious a market premium, that people buying into them did so quite intentionally, conscious of New Urbanist claims. Having bought in believing, among other things, in the place's potential for community, the residents tend to produce and reproduce it through their lifestyles and their accounts thereof. Mike Watkins of DPZ, who worked extensively on the design of the Kentlands, still serves as the Town Architect, and who has lived there since the beginning, says of that project, "the community attracts people who value community, not surprisingly, so these people moved in ready to engage themselves in the lives of their neighbors, in their community and in the city" (Interview with author, December 5, 2005). Richard Arkin, an early Kentlands resident and active citizen leader, remembers, "we propagandized the concept a lot … the virtues of neotraditional design and development and community" (Interview with author, June 19, 2004).

The second factor is that most New Urbanist projects are Common Interest Developments (CIDs). CIDs are more or less synonymous with home owners' associations and participation in them is specified in the deeds of new housing. CIDs are incorporated into most recent suburban development in the United States (McKenzie 1994). They frequently involve active and powerful representative governance structures which, while they may govern a broad range of activities and services within a development, are primarily concerned with the maintenance of property value. Thus, at the very least, the residents of New Urbanist developments are likely to come together – forge "community" – around this common interest.

The point in both these discussions of the portrayal of the suburbs and the professions is not simply that it was possible for the New Urbanism to appeal to homebuyers' desire for community, their sensitivity to sustainability – and the power of the environmental movement – and diverse dissatisfactions in the professions of planning and architecture. Rather, it is that New Urbanists capitalized on these opportunities by relying upon a single unified set of principles that guided transit-oriented, traditional neighborhood development. Both the New Urbanist concept and the places they built early in the movement's history responded comprehensively and concretely to the concerns and interests that these opportunities embodied. By understanding the way in which the concept and the places captured a new set of interests and commitments from within an existing landscape – geographical and social – we can comprehend how the movement accomplished a shift in the production of the built environment.

Meeting the demands of the market: the secondary mortgage market and profitability

Each of the conditions discussed above has functioned simultaneously as opportunity and constraint for the New Urbanism – frustration with suburban traffic and disarray in the planning profession provided openings that could only be exploited by negotiating well-established conventions. By contrast, the private nature of real estate in the United States has functioned almost entirely as an obstacle. Because producing houses for sale is big business, the New Urbanism has had to contend with the underlying structures of capital that make that business possible. This means dealing both with real estate finance's aversion to risk and its approaches for managing risk – in particular the secondary market in real estate finance and real estate investment trusts.

Christopher Leinberger is a successful developer of New Urbanist projects, but is also widely recognized as an expert on real estate finance – he is currently Professor of Practice and founding Director of the Graduate Real Estate Program at the University of Michigan and a Visiting Fellow at the Brookings Institution. His is the most prominent voice on issues of finance within the New Urbanist movement. In his article, "Retrofitting Real Estate Finance: Alternatives to the Nineteen Standard Real Estate Product Types" (2005), Leinberger unpacks the methodology of real estate finance for a design-oriented audience. He captures the New Urbanism's problem perfectly: the "best design ideas are worthless unless money can be raised to build them" (24).

The dominance of the secondary mortgage market – the trade in now-infamous mortgage-backed securities – as the ultimate source of housing finance means that developers "have had to commodify what they build, and this has meant ensuring that each unit of each product type was adequately similar to every other" (Leinberger 2005: 25). Products must closely conform to the criteria assigned to the type category within which they fall to be able to be backed financially, because bankers and analysts have come to specialize in the details of specific product types. Investment is accordingly channeled into the common denominators of post-postwar metropolitan form: shopping centers, regional malls, office and industrial parks, and housing developments with two and three car garages are all known quantities.

"[M]ixed-use, walkable places [do] not conform to the standard product types," and this has steered money away from the New Urbanism because "Wall Street [does] not like experiments" (Leinberger 2005: 26). Put in practical terms, generating financial support for a New Urbanist project requires getting financial institutions to violate protocols that have proven profitable. It is a matter of asking seemingly risk-averse banks to back something comparatively untested, which, even if successful, they will have to carry indefinitely, because it will not be welcome on the secondary market. Much of New Urbanist development to date has depended upon private equity and partnership with local municipalities (Leinberger 2004).

To the degree that the New Urbanists have managed to navigate past this obstacle, it has been because of the demonstrable profitability of some early developments. The Kentlands, in Gaithersburg, Maryland, is the subject of an oft-cited and seemingly indicative study, showing that housing in New Urbanist new towns commands a 12 per cent price premium above that in neighboring conventional subdivisions, holding lot size and other variables constant (Tu and Eppli 1999).[6] Three related studies show more mixed results, but do not dispute the potential profitability of the New Urbanism (Morrow-Jones et al. 2004; Plaut and Boarnet 2003; Song and Knaap 2003).[7] CNU co-founder Elizabeth Moule explained financiers' change in attitude after an early project in Los Angeles was "appraised at the highest dollar per square foot of any building in Southern California":

> And as soon as all the banks found that piece of information out, you can imagine that their interest changed. So we were then able to use not just any old comp [comparable], we able to use the comp to end all comps. The bank that had lent on that project … was calling us, begging us to do more.
>
> Interview with author, January 26, 2006

Conclusion

The story of "Why New Urbanism now?" is a story about the active relationship between the political, economic, cultural, and professional context of suburban development and specific collective efforts to alter that context, anchored in interventions in the built environment. In previous chapters, we have developed a sense of context and the origins of the collective. In this chapter, we have examined how the New Urbanism engaged with and was shaped by the historical moment of its emergence. Locating this emergence in the interaction between the "what" and the "when," the "who" and the "how," is critical.

The early New Urbanists confronted simultaneously a generic American anti-urbanism and a burgeoning discontent with suburban growth and life. They capitalized upon these carefully, reclaiming the "urban" without actually requiring a return to the city, arguing for the possibility of environmental sustainability without undue sacrifice of comfort, and promising walkability and community while de-emphasizing the seemingly necessary correlate, density. Similarly, through form-based coding, New Urbanists made planning New Urbanism easier, without threatening the gate-keeping position of planners, even providing them with a new normative claim – a return to making good places. With architects, they cushioned a restraint on creativity by reconfiguring a familiar design practice, the charrette, to empower early-career professionals and encourage larger-scale thinking and interdisciplinary conversation. Finally, New Urbanists engaged for-profit developers and real estate financiers by pursuing profitability at the same time as principle. The structural conflict between new building types and the secondary mortgage market proved manageable, if New Urbanist projects could make money.

Importantly, all of these New Urbanist interventions functioned in multiple modes simultaneously. They were discursive – "new urban" – and practical – walking, form-based codes, and charrettes. They were socio-cultural

– sustainability and community – and financial – funded by banks and purchased by homebuyers. But most of all, these interventions assembled a group of social puzzles and solved them through the specifics of the production of built form, both the process of its production and the form itself. Urbanity, sustainability, walkability, and community are all anchored in the particular relations of various elements at various scales. Form-based codes put positively useable form first, and charrettes draw on local vernacular knowledge and build agreement into the details of the design. Lastly, New Urbanist projects made profits, because they were places that people wanted to buy.

Revealing this reciprocal relationship between context, collective, and the built environment is at the heart of the project encapsulated by "Why New Urbanism now?". We see not only how this particular movement responded to and changed the context within which it operated, but also how that context both influenced the form the built environment would take and was influenced by it. This last claim is key: that the New Urbanists could alter the social context in which they were operating through interventions anchored in the details of built form and its production shows how the built environment reflects back on the social. It is more than a static container for social life.

Having examined the New Urbanism's emergence and consolidation, we now turn to the question of its endurance. Chapter 5 analyzes how the movement weathered ideological attacks and escaped politicization as it grew and matured.

5 Surviving its critics and reinforcing its position

As the New Urbanism emerged, it confronted criticism and encountered further opportunities to amplify its influence. In this chapter, we will consider how the movement managed attacks on its progress and what measures it took to secure its position. The New Urbanism's pragmatism, comprehensiveness, and concrete qualities enabled it to escape politicization and weather antagonistic professional cultures, while its simultaneously principled and empirical approach made prominent interventions possible.

The New Urbanism's critique of conventional suburban development and its approach to reform have significant ideological implications and both conservative and liberal critics have paid attention to the movement. While this attention constitutes evidence of the significant place the New Urbanism has assumed in the social landscape, it also raised the possibility of concerted opposition that might interfere with the movement's prospects. In fact, the nature of the movement and its products made consistent opposition from established ideological camps in American public discourse difficult. Accordingly, the New Urbanism evaded attempts to identify it with one vein of thought or another in any enduring fashion and proceeded comparatively unmolested. Similarly, while the movement has repeatedly come into conflict with various elements of the architectural profession, its concrete progress has undermined aesthetic complaints and clashes with conventions in the field.

As it survived its critics, the New Urbanism also capitalized on further opportunities to distinguish itself, notably on the national stage. First through shaping the HOPE VI program that rebuilt federally subsidized housing projects, then by its rapid and dynamic response to the destruction of the Gulf Coast by Hurricanes Katrina and Rita, and, more recently, by its proposals for post-earthquake Haiti, the movement garnered important publicity and reinforced its status as a significant player in the broader field of metropolitan policy, planning, and design. Consequently, the New Urbanism not only weathered exposure to ideology and convention, but advanced its project in this period.

The New Urbanism in the ideological press

Determining the ideological valence of the New Urbanist phenomenon as it emerged enables our understanding of what support the movement might draw from what quarters – and what opposition; what opportunities might open to it in which political moments – and which might close; and, fundamentally, how the New Urbanism engages the ideologically fraught intersection of politics and economics in the United States. My data, for this analysis, is drawn from a comprehensive search of prominent "ideologically identified" publications for mentions of the New Urbanism. I characterize these publications as ideologically identified because they take positions on current events that are conspicuously politically non-neutral and are also either explicitly affiliated with a particular ideological stream – as explained, for instance, in their mission statements – or are commonly known to be associated with such a stream. Broadly, these ideologies can be characterized as either conservative or liberal, but there are significant distinctions among subcategories of each.

Generally speaking, as we will see below, conservative critics[1] are primarily opposed to both the New Urbanism's underlying claims about conventional suburban development and the movement's efforts to build, but they cannot ignore completely the appeal of the New Urbanism's seeming traditionalism. On the liberal end,[2] the response is, interestingly, also predominantly one of opposition to both New Urbanist critique and New Urbanist construction, but is also similarly undercut by the admission that the New Urbanism is an improvement on the much reviled – among liberals – suburban status quo.

The conservative response

Libertarian critics offer the most intense and coherent opposition to the New Urbanism. They represent the movement as a burgeoning, top-down, federal governmental conspiracy to force Americans to live in compactness equivalent to Manhattan and mandate the replacement of their cars with expensive light rail. The response from mainstream conservative critics, on the other hand, ranges from laudatory to condemnatory. The New Urbanism is by turns touted as a traditionalist solution and attacked as elitist and authoritarian. The response of the near-right tends to focus on the question of choice and to reassert the apparently overwhelming preference of the American populace for conventional suburbia and automobility. It incorporates a suspicion of expertise and government and some mild accusations of elitism. Yet the near-right response also includes some rather neutral reportage of research on the financial success of New Urban developments, presumably compelling to a pro-capitalist readership.

Conservative criticism of the New Urbanism focused on four themes – coercion, elitism, preference for sprawl, and traditionalism. Each might have formed the basis of a coherent response that would have threatened the New Urbanism's success.

The most radical and intense complaint about the New Urbanism among conservatives is that it is, or will become, mandatory through its embodiment in government policy. This complaint comes particularly from libertarian critics. The concern takes the form of the suggestion that either the New Urbanism has been forced on people by governments "infested" (O'Toole 1999) with New Urbanist sympathizers or that the movement will subvert democratic processes to accomplish its goals.

Randall O'Toole writes:

> I have often said that I would not object to New Urbanism if it were voluntary. … When I bicycle by New Urban neighborhoods in Portland, Denver, or elsewhere, I get a sick feeling in my stomach because I know many of the people in these neighborhoods have been forced to live there because of various government coercions, and that this has severely degraded their quality of life.

Samuel Staley, of the libertarian Reason Foundation, claims the New Urbanists "are calling for a radical increase in government control over land development and, ultimately, over the physical dimensions of neighborhoods" (2001). Fred Barnes, editor of the near-right *Weekly Standard* writes, "It's the anti-sprawl movement that wants to force a lifestyle and a housing pattern on unwilling Americans. For activists like [CNU founder Andres] Duany, democracy is often an impediment" (2000). Finally, Stephen Hayward writes in *Policy Review* that, "They approach urban problems with an attitude that could justly be described as 'Planning Über Alles'" (Hayward 1998).

This concern brings out the tension that some conservatives perceive between freedom and government. Yet this perception is the most intense among libertarians and has seemed overstated to the more mainstream conservatives. Libertarian critics concerned about coercion have undermined their position by failing to provide clear evidence that coercion actually occurs.

A critical claim with broader appeal among conservatives is the charge of elitism. Some conservatives have argued that anti-sprawl movements, including the New Urbanism, are primarily venues for the aestheticist, urbanite elite. Robert Bruegmann is the author of the recent *Sprawl: A Compact History* (2005), in which he seeks to reclaim the term sprawl. Discussing his book in *The American Enterprise*, he suggests that the anti-sprawl movements are fundamentally elitist: "Sprawl is where other people live, particularly people with less good taste" (Bruegmann 2006). Bruegmann continues, "Class-based aesthetic objections to sprawl have always been the most important force motivating critics" (ibid.). Similarly, Hayward, cited above, lumps the New Urbanism with other liberal reforms and complains that they merely represent: "Elite distaste for suburban life" (Hayward 1999).

The editor of *The American Enterprise*, Karl Zinsmeister, associates the "organized disdain toward today's horizontal communities" with the self-interest of current suburbanites: "'I got mine, now pull up the ladder' snobs are a dime a dozen in these movements … " (2006). In the same issue of *The American Enterprise*, the sometimes New Urbanism-sympathetic Catesby Leigh attributes a related attitude to the New Urbanism: "[New Urbanist traditionalism] ought to have attracted the support of many cultural conservatives. Instead, the movement is weighed down with an image of control-freak elitism" (2006b).

Conservative accusations of elitism in the New Urbanism are tied up with the effort to consistently oppose liberal depredations on the culture of the American middle. They use the similarities between traditional liberal and New Urbanist attitudes toward suburbia to collapse the two groups together. This effort is consistent with broader conservative efforts to define "elite" in cultural terms, foregrounding a kind of anti-intellectualism that disables reasoned critique.

Nevertheless, the conservative accusation of elitism has faltered due to the disjuncture between the essence of the criticism and the material aspect of the movement as a specific product on the residential real estate market. Regardless of the movement's aspirations, to date its products appeal primarily to upper-middle class homebuyers, a group comparatively unsusceptible to anti-elitist arguments. This anti-elitist critique of an upper-middle-class phenomenon resembles the postwar, elite liberal response to the suburban boom: opposition to the building of Levittowns on behalf of suburbanizing masses who, themselves, eagerly snatched them up. And its effect on the behavior of homebuyers is similarly minimal.

Conservative critics have also suggested that New Urbanist reforms are unnecessary or redundant because the market already responds to homebuyers' preferences – the proof is in the suburban pudding: Americans love sprawl. USC economist Peter Gordon is foremost among the defenders of sprawl. Anti-sprawl movements are mistaken, he claims, in an interview in *Reason* magazine because,

> People are getting the neighborhoods they want … . That's the weak link in the New Urbanism. If there were a grain of truth in their view, we would soon see people demanding it, and developers would strive to provide it. Builders are not ideologues.
>
> Henderson and Moore 1998

Fred Barnes writes similarly in the *Weekly Standard*, "For all the scorn that's heaped on the suburbs – and especially on subdivisions of nearly identical houses on the fringe of metropolitan areas – people like living there" (2000). Of course, if this were the case, New Urbanist developments would find no buyers, but in fact they do sell.

The primary reason that conservatives have failed to become a substantial obstacle for the New Urbanism arises from the fact that they find quite a bit to admire about it. The movement's reference to traditional small towns in particular appeals to their commitment to traditional values. In a 1996 issue of *The American Enterprise*, for example, editor Karl Zinsmeister writes:

> The kinds of traditional towns and neighborhoods we once had all across this country, the very locales where our own parents grew up, were mostly excellent places to call home. They captured the wisdom of centuries of human experience at living together. All we need to do is return to their principles and rules – and real communities can flourish again.
>
> Zinsmeister 1996

Furthermore, Zinsmeister contradicts Gordon's claim that the market satisfies demand, concluding, "People accept suburbs, but they aren't particularly enthusiastic about them" (ibid.). He is therefore pleased to recognize New Urbanism's potential as a traditional alternative. This is reinforced by Philip Langdon's finding, in the same issue, that, "Traditional community development also seems to accord with a broader traditionalism in values" (Langdon 1996).

Conservative critics, in other words, cannot resist the note of traditionalism in the New Urbanism.[3] They uncritically accept the connection between tradition in planning and design and traditionalism in values, a connection inherent in the New Urbanism's claims about the ties between built form and behavior. These sympathetic accounts of the movement come from one particular issue of *The American Enterprise* dedicated to the subject and entitled, "Cures for Lonely Suburbs and Dying Cities" (November/December 1996). The centerpiece of the issue was an unwavering critique of conventional suburbia by Zinsmeister. Most of the appeal to conservative critics arose from three essential elements of the New Urbanist approach: its respect for traditional design; its deference to the private market; and, its challenge to zoning – long viewed by some conservatives as unjust, government "takings" of their property.

To confirm the New Urbanism's appeal among conservatives, we should also note that, in the aftermath of Hurricanes Katrina and Rita, the Republican governor of Mississippi, Haley Barbour, quickly took up the New Urbanism and New Urbanists as the solution to rebuilding the devastated Gulf Coast, an issue to which we will return below.

And this is the real point: conservatives have not mounted a sustained and uniform challenge to the progress of New Urbanist development, because they cannot maintain a consistent ideological position vis-à-vis the New Urbanism. The New Urbanism avoided consistent characterization and, thus, conservative opposition, because it has been pragmatic in its approach to an integrated and comprehensive re-conception of real estate development that does not highlight defining principles that fit neatly with pre-existing ideological tendencies. New Urbanists have not challenged the profit motive in real estate development, nor have they oversimplified the complexity of the political, cultural, regulatory, and financial processes and relationships in suburban development. While avoiding coherent opposition from conservative critics, they have simultaneously caught the attention of homebuyers, generating a significant demand for their product.

The liberal response

Critics writing in liberal publications have found many opportunities to engage with the New Urbanism, and much of that engagement has been antagonistic. There are three primary complaints: that the movement, in its ambition, has come to serve the Republican forces of evil; that the focus on design entails the commodification of community and the de-politicization of the development process; and, that the rhetoric of community taken up by the movement has historically been used – and will continue to be used – to exclude and deprive.

Nevertheless, the New Urbanism's treatment in the liberal press has not been uniformly negative. Many of the most strenuous complaints have been most clearly articulated in publications identified with the far-left and, thus, have gained less traction or support within a broader liberal group. More importantly, some liberal critics find aspects of the movement to admire. Accordingly, the New Urbanism has weathered its exposure to liberal ideology in much the same way it did that of the conservatives.

The most pointed liberal attack on the New Urbanism has been the accusation that the movement is exclusive, a virtually gated project. The movement's efforts to suggest that their designs promote "community" have inspired liberals to unpack the exclusionary history of the vaunted concept. Geographer and social theorist David Harvey's article, "The New Urbanism and the Communitarian Trap," is the most powerful expression of this concern and is frequently referred to in other liberal critics' writings (1997). Harvey forcefully questions what the New Urbanism means by community. He warns, "'Community' has ever been one of the key sites of social control and surveillance, bordering on overt social repression" (ibid.: 3). Not only have the foundations of community often been predicated on exclusion, but a major reason for the historical out-migration of villagers is that village communities were "oppressive to the human spirit and otiose as a form of sociopolitical organization" (ibid.).

Urban historian Thomas Bender sounds the same call when he discusses the New Urbanism in his review in *The Nation* (2003) of Vincent Scully's recent *Modern Architecture and Other Essays* (2003). Bender characterizes the New Urbanism as offering "city-lite" and echoes Harvey in cautioning that community " ... carries its own burden. The quest for community, almost by definition, is the pursuit of sameness, a sense of we-ness. And 'we' implies an excluded 'them'" (Bender 2003: 40). For these critics, the New Urbanism's promise of community raises fears of exclusion, repression, and the reinforcement of privilege.

The negative repercussions of the pursuit of community, however, speak powerfully only to a few. Community remains a prevalent ideal among Americans, including mainstream liberals. While few liberals would contest the legitimacy of Harvey's or Bender's criticism, nor do many take up the critique of community as a rallying cry. It has too positive a connotation to be subjected to systematic dissection, especially for the homebuying public. Furthermore, once the community concept has been emphasized in marketing New Urbanist developments, it takes on a self-fulfilling character: residents who bought with community in mind both enact it and perceive it.

Related to concerns about community, the New Urbanism has had to endure liberal claims that it represents nostalgia for a world far more unequal than the current era. Nostalgia has the potential to be reactionary or politically retrograde. Thus, architectural historian Witold Rybczynski, writing in *The New Republic*, complains: " ... in practice, it's difficult to disentangle what we admire about the past from what we would rather forget – and almost impossible to re-create that partly imagined past" (1995). He explains that the appeal of the New Urbanism

> ... is as much to a fantasy of a more ordered and harmonious past as it is to some vestigial memory of village democracy. ... Like the campaign for family values, traditional neighborhood development is an attempt to deal with difficult modern conditions by invoking, uncritically, a return to 'simpler' virtues.
>
> Rybczynski 1995

The New Urbanism's alleged nostalgia is at odds with core liberal values – commitments to progress, in particular, and equity. Nostalgia is dangerous because it elides or obscures aspects of the past we would rather forget. Nostalgia is an exercise of the imagination and necessarily an avoidance of reality. Still, this accusation has had minimal impact for two reasons: first, nostalgia is only implicit in New Urbanist discourse, if present at all, and is not an articulated principle of the New Urbanist approach. New Urbanist designers explain historicist vernacular architecture with technical rationales deriving from the analysis of spatial dynamics, as opposed to a yearning for some simpler time.

Second, as Mark Greif argues in *The American Prospect*, those most guilty of nostalgia may be homebuyers, not designers or developers. He writes, "A bit of marketing experience quickly added the further rule of historic recreation. If it looked old, it sold. To deliver old-fashioned community, planners had to make good on some heavy-duty nostalgia" (Greif 2000). This puts liberal critics in the same position as the conservatives who target elitism, attacking something a significant proportion of their audience is practicing unselfconsciously – and reminds us of the New Urbanist penchant for pragmatic compromise.

A final liberal criticism of the New Urbanism is that it is driven by an overwhelming desire for financial success. Critics taking up this complaint express profound doubts about New Urbanism's alliance with for-profit developers. The movement seems amorally apolitical, climbing into bed with whomever will help it get where it wants to go. More, because of the political economic structure of real estate development, this position puts profit before questions of good policy. Daniel Lazare writes in *In These Times* that the greatest problem with the New Urbanism is "its rejection of politics," doing nothing to reverse "the great wave of de-urbanization that ... dispers[ed] workers, weaken[ed] working class consciousness and diffus[ed] class struggle" (Lazare 1997).

This sort of criticism has been most forcefully lodged against the New Urbanism in light of the CNU's participation in the rebuilding of the post-hurricane Gulf Coast. Thus, Mike Davis, writing in *Mother Jones* about the situation in Louisiana and Mississippi after Hurricanes Katrina and Rita in 2005, presents the New Urbanists as over-reaching dupes caught up in Republicans' sinister machinations, and detects a dangerous naïveté (Davis 2005). Unprincipled ambition has gotten them into serious trouble, according to Davis, starting with the legitimating function they played in the planning process for the Mississippi Gulf Coast. He says about the CNU-led, week-long, design marathon Mississippi Renewal Forum:

> In the end, however, what was important was not the actual content of the charrette, nor the genuine idealism of many participants, but simply the legitimacy and publicity that CNU gave to [Mississippi governor] Barbour's agenda. Duany, who never misses an opportunity to push his panaceas to those in power, has foolishly made himself an accomplice to the Republicans' evil social experiment on the Gulf Coast.
>
> Ibid.

According to these critics, because the New Urbanism does not problematize the market, it falls into the reproduction of other social ills – class and race inequality, in particular. Yet, as a criticism likely to undermine the New Urbanism's progress, this falls short because participation in American capitalism is hardly a damning charge in the current era, even among liberals. Moreover, it is easy to see a different interpretation of the events these critics portray as service to the "Republican's evil social experiment." The story of the New Urbanism's involvement in planning the reconstruction of the Gulf Coast could be read as one about pragmatism, non-partisanship, and accomplishment: the movement achieved something affirmative in a difficult situation that could ultimately benefit hundreds of thousands of people, whatever their political persuasion or that of their elected officials.

Finally, the New Urbanism has survived exposure to the liberal media by inspiring some positive acclaim. While liberal critics are sometimes cynical about New Urbanism's perceived nostalgia and skeptical of the movement's claims to recover community by design, some mainstream liberals appreciate its accomplishments and its improvements over conventional suburban development. Near-left liberal critics admire the New Urbanism's demonstrated success with homebuyers (see Morrow-Jones et al. 2004; Plaut and Boarnet 2003; Song and Knaap 2003; Tu and Eppli 2001, 1999).

Sympathy for the New Urbanist project can be found even among some of its most ardent detractors. Even David Harvey begins his biting critique by saying:

> There is much in this movement to commend it ... first, the willingness to think about the place of particular developments within the region as a whole and to pursue a much more organic, holistic ideal of what cities and regions might be about ...
>
> Harvey 1997

And Mark Greif writes, "At their best – and their best could be very, very good – the New Urbanists bundled insights about Main Street and traditional, dense neighborhoods into planning principles for revitalizing all kinds of anonymous public spaces" (Greif 2000).

It is writers on the near-left, however, who find the most to praise in the New Urbanism, focusing on its pragmatism and reformism. In his 1995 article in *The New Republic* subtitled "The Rise of Family Values Architecture," Witold Rybczynski notes with approval the positive response to the New Urbanism in the broader public, "Traditional neighborhood development is about to become the hottest new idea in suburban planning" (Rybczynski 1995). He suggests that the popular response is both normative and, at least in part, correct, "And the public seem[s] to be onto something that the critics, self-styled champions of the avant-garde, [are] not. It recognize[s] a quality that [has] all but disappeared from American town planning: beauty" (ibid.).

Later, Rybczynski, as *Slate*'s architecture critic, went further in his account of the New Urbanism's role in the rebuilding of the Gulf Coast. In an article entitled "How To Rebuild New Orleans: Start with a visit to Denver" (2006),

Rybczynski focuses on a particular project's success as a way of defending the New Urbanism's potential role in rebuilding New Orleans. The article is clearly a response to the attacks New Urbanists had been fending off, particularly from Tulane architecture professor Reed Kroloff. Kroloff argued that a New Urbanist New Orleans will "look like your grandmother's hometown" and that New Urbanist planning will not serve ordinary people (from the *Milwaukee Journal Sentinel*, cited in Rybczynski 2006). Noting that his grandmother "grew upon in downtown Lwow," Rybczynski responds to Kroloff with a description of the Stapleton project in Denver, Colorado, with a super-center – including Wal-Mart, Sam's Club, and Home Depot – on its edge: "This is not doctrinaire new urbanism, but in the real world, planning doctrines must adjust" (ibid.). Rybczynski goes on to tout the pleasant, generic, modern feel of some of the development and the historicist appeal to homebuyers in the rest of it. He concludes that Stapleton is an example of getting something pretty good done and, thus, that CNU might have the ability to realize something for the residents of the disaster areas incorporating at least the virtue of completeness.

Finally, Jennifer Bradley, writing in *The American Prospect*, similarly applauds the New Urbanism's pragmatism (2000). She writes that, despite New Urbanists' "urbanity,"

> They recognize … that people want new houses in new developments, away from the real and perceived problems that are built into old houses in old neighborhoods, and they conclude that since new development is going to happen anyway, they must "reshape new growth into the most benevolent form possible."
>
> Ibid.

In the aggregate, then, the liberal response to the New Urbanism has not interfered with its progress, because liberals, as a group, have never overcome a sense of ambivalence about the movement. In fact, significant local obstruction of New Urbanist projects on a liberal basis is neither reported in the liberal press nor in New Urbanist publications. Instead, New Urbanist places themselves have largely elicited sympathetic responses from mainstream liberal critics. This is because these places address many traditionally liberal criticisms of suburbia. In sum, the source of liberal ambiguity about the movement is a confusion about whether the New Urbanism is the right answer or just a superficially appealing one dependent upon regressive means – a traditional problem of reform vs. revolution. This debate affords the movement substantial free publicity, this time among a liberal readership.

Conservative critics have tended to characterize the New Urbanism as coercive, elitist, or unnecessary. Liberal critics have leveled accusations of exclusivity, nostalgia, and collaborationism. While these complaints have focused criticism, none has galvanized any coherent response from within either ideological group. In the cases of both groups, the sharpest attacks have come from the most ideologically extreme and this partly explains their failure to catch on. More importantly, both conservatives and liberals in the aggregate are ultimately ambivalent about the New Urbanism. While they are critical, conservatives also find the movement's traditionalism appealing, while liberals also appreciate it as an improvement upon the suburban status quo.

The rise of the New Urbanism demonstrates that the progress of a movement and its products can be as much a matter of evasion as anything else. The New Urbanism escaped the critical grasp of the ideologically identified press because of its comprehensiveness and pragmatism, but, as importantly, because the places themselves confound ideological critics of various stripes. Making places is the apotheosis of pragmatism: once on the ground, new neighborhoods function substantially outside of the discursive realm over which the critics exercise influence. The complexity of the movement's places and rhetoric made it impossible for ideological groups to isolate a coherent and ideologically consistent central core of commitments by which to define the movement. This is not because the New Urbanism conceals its principles – it broadcasts them – but rather that no one principle in the integrated approach could be demonstrated to be materially more essential than another.

Comprehensiveness and pragmatism are integral parts of the New Urbanism, but they resulted substantially from the effort to build within a landscape of constraints and opportunities. The most important of those constraints, the private market in real estate, demanded that the movement incorporate the assumptions of development finance and the predilections of homebuyers into their approach for them to be able to build on any scale. The movement could build, because it recognized that the reform of suburban development had to be reform for sale. They had to appeal to the homebuying public, liberal and conservative.

The contours of the New Urbanism as described by architects and architecture critics

As I argued in Chapters 2 and 4, the social dynamics within the professions of architecture and planning are critical to understanding the rise of the New Urbanism. While it has not gone without comment (see, in particular, Talen 2005), planners absorbed New Urbanist claims into their practice without too dramatic a public discussion. Likely because it conflicts more directly with core commitments of contemporary architecture, the New Urbanism has caused much greater consternation in that field.

The published criticism of the New Urbanism by architectural academics and professional critics highlights the subjects on which the movement and the architectural establishment have come into conflict. This tells us something not only about the New Urbanism – particularly about its position on style and its pragmatism about development – but also about the field of architecture itself, what it values. Contemporary architecture is still heavily influenced by early mid-twentieth century Modernism. The result, despite Mies Van Der Rohe's declaration that "form follows function," is an emphasis on a particular kind of style, complicated somewhat by the rise of various post-Modernisms (see Hirt 2009 for an interesting discussion of New Urbanism as post-modernist).

As we have seen, the compromise between this residual Modernism and more recent challenges has come together in the conception of the architect as artist, freely exercising creativity in the creation of a building as sculptural object, an abstract response to contemporary urban conditions (Sarfatti Larson 1993, Blau 1984, Cuff 1991). New Urbanism, by contrast, while not settled on aesthetic specifics, does subordinate style to function, both in the design process and in the final product. Furthermore, the artistic ideal in architecture also resists the need to bend creativity to the requirements of developers or the market, often putting pure invention ahead of profitable implementation. The New Urbanism, again by contrast, is attempting to change the way middle America lives and thus privileges the pragmatic in producing places that middle America will prefer to sprawl, getting them done and lived in. This willingness to engage pragmatic concerns, like a readiness to make style secondary – or even pander to popular tastes – defines the key conflicts between the New Urbanism and the public voices of the architectural field.

Nostalgia or precedent

The most common reaction to New Urbanism among architects is an attempt to dismiss it as nostalgic. Nostalgia is not a style, not a legitimate position in the ongoing debate over style among those entitled to advocate one position or another. Nostalgia is a sentiment. Worse, it is a marketable, low-culture sentiment, retrograde. Herbert Muschamp, then architecture critic for the *New York Times*, once referred to "the oxymoronically named New Urbanism movement, with its compact suburban subdivisions and houses fabricated from Cheez Whiz" (Muschamp 2000) and Reed Kroloff, dean of Tulane's architecture program, calls it "treacly, sugarcoated, neo-precious" (as quoted in Dellinger 2006). It is either too processed or too sweet.

The New Urbanism is, of course, backward-looking in its reliance on precedent in spatial relations. That this is confused with a romanticization of the past, a past which most critics go on to assert is imaginary, is problematic for the New Urbanism because of its assumed political content – we have just discussed conservatives making the same conflation. The most widely cited attack on the New Urbanism in the post-Katrina period was that of Eric Owen Moss, architect and director of the Southern California Institute of Architecture (SCIArc). In *The Washington Post*, Linda Hales quoted Moss as saying about New Urbanist plans for the Mississippi coast, "The ideological image-making would appeal to a kind of anachronistic Mississippi that yearns for the good old days of the Old South as slow and balanced and pleasing and breezy, and each person knew his or her role" (Hales 2006). As Blair Kamin, Pulitzer Prize-winning architecture critic for *The Chicago Tribune*, wrote, "Moss didn't say that Andres Duany and Elizabeth Plater-Zyberk wanted to bring back the Jim Crow laws, but he might as well have" (Kamin 2006). Mike Davis wrote similarly in *Mother Jones*, "Duany, as usual, whipped up a revivalistic fervor that must have been pleasing to Barbour and other descendants of the slave masters ... " (Davis 2005). The confusion of precedent and nostalgia opens the New Urbanism up to accusations of regressive politics from a generally progressive design field.

Pragmatism

The New Urbanists are frequently accused of collaboration with development interests and government. Christopher Hawthorne wrote, in *The Los Angeles Times*, that they are "too friendly to developers" according to their academic and avant-garde colleagues, and:

> The New Urbanists ... [unlike "celebrity designers" who "have failed to find influence in government"] have been more skilled at making themselves welcome on Main Street and in the corridors of power – even as their stock among fellow architects, particularly young and urban ones, has plummeted.
>
> Hawthorne 2005

The New Urbanists require the cooperation of many to accomplish their goals, including developers and government, and are generally unashamedly pragmatic about it. Another of Hawthorne's analyses gets straight to the most common motivation for the accusation of collaboration:

> It is this very attempt – to lay traditionalism like a veneer over directly pro-business policymaking – that opens up the New Urbanists to an oft-heard charge: namely, that they are willing co-conspirators in a kind of nature-gobbling development that is significantly less enlightened than it claims to be.
>
> Ibid.

Reed Kroloff makes a similar claim in a profile of Duany in *The New York Times* (Pogrebin 2006). Davis, in his *Mother Jones* piece, accuses Duany of lending publicity and legitimacy to "the Republicans' evil social experiment on the Gulf Coast" (Davis 2005). So, New Urbanism is not simply collaborating, but is being co-opted to serve the forces of evil, i.e. capitalism and Republicanism.

All of this makes very clear that the practice of the New Urbanism involves the pragmatic compromise of cooperation between designers, developers, and government, but also indicates that this cooperation is a violation of the design field's principles. In the end, this tells us more about architecture than New Urbanism: the ongoing commitment to the ideal of architect-as-artist means their creative genius should not be hindered by practical concerns.

Held to an unreasonable standard

A final criticism of the New Urbanism is that they claim to do too much and accomplish too little. In 1996, Muschamp gave the New Urbanism its ultimate press clipping. He wrote, "The Congress for the New Urbanism is the most important phenomenon to emerge in American architecture in the post-Cold War era" (Muschamp 1996). He goes on to say, though, "Like the modernists, the new urbanists rely too much on esthetic solutions to the social problems created by urban sprawl" (ibid.). In the same year, Alex Marshall wrote similarly about the New Urbanism in *The Washington Post*: "There's only one problem: New Urbanism doesn't work. Its proponents are selling something they can't deliver without charging a far higher price, and without making changes far more fundamental than redesigning a few homes" (Marshall 1996). Marshall concludes:

> [New Urbanists] have to get their priorities straight. They have to give up the dollars generated by alliances with home builders intent on moving development ever outward. New Urbanism's contribution to city planning will remain almost purely stylistic, unless it makes more effort to change the basic pattern of suburban development.
>
> Ibid.

This criticism both ignores the movement's involvement in HOPE VI and other urban infill and begs the question of how a group of designers and planners is supposed to produce affordable housing in a strictly private market.

Yet many of the New Urbanism's most ferocious critics recognize, and this may be the reason for some of their ferocity, that by the fall of 2005 and the most urgent moment of need for large-scale and rapid urban design, the New Urbanism was "the only game in town." The sympathetic Hawthorne puts it more elegantly, referring to the Democratic defeat in the 2004 Presidential election:

> The Biloxi charrette, in other words, may go down as the architectural elite's Ohio: the place it watched rather helplessly as its ideological opponents outclassed it not through nimble thinking or grand theory or inspiring plans but simply by being more disciplined and better organized.
>
> Hawthorne 2005

New Urbanism vs. Landscape Urbanism

The reproduction in public and professional discourse of the New Urbanism's significance, even primacy, has been reinvigorated recently by the rise of "Landscape Urbanism." Landscape urbanism begins from ecological principles – the natural courses of waterways and distribution of wetlands, the interaction of plant species with subsoil chemical residues from industry, the relation of a site to the sun and wind – and derives the appropriate shape of the city, reconfiguring the built environment to respect these ecological factors and inserting humans into "these flows." It conceives of the city as built up, in the first instance, of ecological relationships (Waldheim 2010).

Landscape urbanism puts landscape architecture at the center of the urban design process. According to architect and landscape architect Mark Rios, "Architects are trained to design objects. They go through design school looking at form and program. Landscape architects look at voids, space, systems, based in training in ecology. They deal with bringing spaces together – how they are transformed through ecology" (Quoted in Birnbaum 2010). This focus on "voids" and "space" enables the privileging of the ecological, because the built environment is not the first consideration. Leading landscape urbanist Charles Waldheim says, "The landscape architect has emerged as the urbanist of our age" (Waldheim 2010). And the landscape urbanism has emerged as the movement of the moment.

Landscape urbanists leverage a general interest in the ecological with a realism about post-industrial urban brownfields and greyfields and their ecological state. More than that, they engage the particular political-economic reality of the second decade of the twenty-first century: neo-liberalism and recession. According to Waldheim, because neither the funding nor public will is available to plan good spaces, landscape urbanism circumvents the planning bureaucracy and proceeds incrementally "to steer or predict future succession using modest resources initially" (ibid.).

Landscape urbanism's rise is important to the New Urbanism because it "was developed in some ways as a direct critique of New Urbanism" (Waldheim quoted in Stein 2010) and, more, "to provide an intellectual and practical alternative to the hegemony of the New Urbanism" (Waldheim 2010). Not only does "hegemony" attribute a great deal of importance to the New Urbanism in urban design, but the conceptual opposition of the two urbanisms has inspired significant discussion and debate that ranges from the popular press to the design professional blogosphere, returning the New Urbanism to consideration again and again. The New Urbanists have not just basked in the perverse glow of the landscape urbanist challenge either, but have engaged it actively. Charles Waldheim was invited to speak at one of the large sessions at CNU 19 in Madison, Wisconsin, in 2011, and Andres Duany has been a prominent player in "LU vs. NU" debates online.

The outcome of this back and forth, while interesting if unknowable as of this writing, is less important than the simple fact of it. However we might evaluate the truth of Waldheim's accusation of hegemony, the New Urbanism has found its way back into a spotlight, courtesy of these debates.

Prominent interventions: the New Urbanism on the national stage

The Congress for the New Urbanism and the movement more generally played central roles in several national and international events, giving them a degree of attention that they have otherwise struggled to attain. In at least the sense that "no publicity is bad publicity," these interventions have been successful by requiring bureaucrats, policy-makers, researchers, and the press to engage the New Urbanism as a set of ideas. Architect and professor of architectural design Mike Pyatok captures the conundrum of the New Urbanist pursuit of an enhanced public profile:

> The CNU, whether consciously or not, wants larger projects to achieve real impact and to provide demonstrations it thinks are worth emulating. So the CNU, with only a few exceptions, gravitates toward larger-scale sponsors. In doing so, it often finds itself having to represent powerful interests that are displacing others in the way of development … .
>
> Pyatok 2000: 806

It is an open question how much New Urbanist involvement has materially changed the outcomes of these programs and events. Nevertheless, the response to Katrina and Rita, foremost, but the HOPE VI redevelopment projects too, did perhaps more than anything else to bring the movement's capacity and accomplishments to a larger public consciousness.

HOPE VI

The HOPE VI program – Housing Opportunities for People Everywhere – emerged as a program originally called the Urban Revitalization Demonstration in 1992 in response to a study by the National Commission on Severely Distressed Public Housing which identified 6 per cent or 86,000 units of the nation's public housing as uninhabitable (HUD nd; Popkin et al. 2004: 20). The program was first funded by congressional appropriation in 1993 and has since made more than $6.7 billion in grants (HUD nd). HOPE VI's purpose is to replace distressed public housing with redesigned, mixed-income projects, provide rental vouchers for former residents, and make supportive services available to the inhabitants of the new housing.

HOPE VI is relevant here because of its incorporation of New Urbanist principles beginning in the mid-1990s. The US Department of Housing and Urban Development's (HUD) guidelines for HOPE VI redevelopment projects encourage mixed-use, walkability, traditional architecture, and integration with surrounding neighborhoods through continuous street grids (as opposed to modernist "super-blocks"), among other things, in response to concerted New Urbanist intervention. This influential role for the CNU was consecrated in 1996 when HUD Secretary Henry Cisneros signed the CNU Charter at the Charleston Congress. In 2000, HUD and the CNU co-published "Principles for Inner City Neighborhood Design: HOPE VI and the New Urbanism" "intended to serve as a framework for both the design and the process of designing HOPE VI" and reasserting the strength of the relationship under Secretary Andrew Cuomo (HUD and CNU 2000).

Accordingly, CNU and the New Urbanism are closely associated with what researchers from the Urban Institute and the Brookings Institution call "a dramatic turnaround in public housing policy and one of the most ambitious urban redevelopment efforts in the nation's history" (Popkin et al. 2004: 1). Similarly, geographer James Hanlon writes:

> No other federal housing policy initiative in recent memory has maintained a higher profile than … HOPE VI. … [It] is held in high regard by public officials, policy makers, and housing professionals, not least because the country's most troubled public housing projects have been razed, and once neglected and downtrodden neighborhoods are now infused with a renewed sense of optimism.
>
> Hanlon 2010: 80

The evaluations of HOPE VI-funded redevelopment projects are mixed. The Urban Institute and Brookings Institution work mentioned above is a meta-study, aggregating research on HOPE VI to 2004. They find some evidence that the New Urbanist influence may have had good effects:

> Four types of positive physical changes were common to all the sites in the Interim Assessment study [referring to a 2003 evaluation commissioned by HUD]: an overall reduction in density; connecting properties to the surrounding area through the introduction of sidewalks and street grids; physical changes that increased safety, such as private entrances that face the street; and improved exteriors—for example, bay windows, front porches, or gabled roofs.
>
> Popkin et al. 2004: 20

Hanlon's review of the literature published six years later suggests findings of "substantial improvements in poverty rates, household incomes, property values, and mortgage investments" associated with HOPE VI, but reinforces some of the hesitations that Popkin et al. express (Hanlon 2010: 81). Among other difficulties, many of the measured benefits of HOPE VI redevelopment for the former residents of the "severely distressed public housing" in question come primarily from the demolition of projects that had become the residence of last resort for the most severely disadvantaged (Popkin et al 2004; Hanlon 2010). The program was virtually guaranteed to improve concentration of poverty statistics, the prime example, just by displacing extremely poor residents and scattering them in relatively less disadvantaged neighborhoods.

The evidence suggests that New Urbanists have contributed importantly to a prominent federal housing program, but the ramifications of that association extend to the shortcomings of that program (echoing Pyatok). A few changes to federal housing policy were critical to make HOPE VI possible and they have invited much criticism. First, HUD had to eliminate a "one-for-one" housing unit replacement requirement to accommodate smaller scale, less dense projects (Hanlon 2010: 83). Second, they needed to ease eligibility requirements to allow local housing authorities to attract more middle-income residents to dilute the poverty of the projects (ibid.). Third, HUD changed development financing rules so that HOPE VI grants could leverage significant additional funds, bringing private developers, non-profits, and non-housing authority state actors into the construction of mixed-income projects (ibid.). Accordingly, many of the units that were destroyed will not be replaced; among those that have been, there are fewer available for the very poor; and, the profit motive has taken over from the social safety net as their justification (Popkin et al. 2004). Furthermore, while HOPE VI funded demolition tends to occur quickly, the construction of new units is often quite slow (ibid.).

New Urbanists cannot be held accountable for these particular failures of the HOPE VI program, but they are frequently tarred with the same brush. Hanlon, for example, portrays the New Urbanists as facilitating the "neoliberal transformation of public housing" and writes that the "overtly aestheticized New Urbanist transformation of public housing confers legitimacy upon that transformation while eliding its social and economic costs" (Hanlon 2010: 95). Elliot, Gotham, and Milligan find similarly that the New Urbanism is applied as a framing device by various sides in a struggle over HOPE VI, but that it can ultimately be used to justify even interventions that violate the movement's principles, like the inclusion of big-box stores in a project (Elliot, Gotham, and Milligan 2004). At the same time, and perhaps legitimating New Urbanist involvement, these scholars highlight, and the majority of published research on HOPE VI makes some mention of, the New Urbanism and the movement's association with the program.

CNU responds to disaster: the Mississippi Renewal Forum, Katrina Cottages, and Haiti

Hurricane Katrina hit the Mississippi Gulf Coast in late August, 2005, followed shortly by Hurricane Rita. Katrina, in particular, devastated much of an eighty-mile stretch of coastline by washing houses, roads, and bridges away in a storm surge – 55,000 homes "obliterated" and 80,000 damaged, according to Stark (2006). Less than two weeks later, Mississippi architect and CNU member Michael Barranco contacted Leland Speed, a retired real-estate investor. Before the hurricane struck, Speed had been serving as the Executive Director of the Mississippi Development Authority, the state's "lead economic and community development agency" whose mission is "to foster a strong state economy and vibrant communities through innovation, use of talent and resources to improve our citizens' lives."[4] Barranco sought out Speed because it was common knowledge among local New Urbanists that he "had read and admired [Duany, Plater-Zyberk, and Speck's] 'Suburban Nation'" (Hawthorne 2005). Speed, moreover, had the ear of Jim Barksdale, former Netscape CEO and Governor Haley Barbour's choice to chair the Governor's Commission on Recovery, Rebuilding and Renewal – established in late 2005 to oversee the post-hurricane recovery effort.

Shortly after Barranco contacted Speed, Andres Duany flew from Miami to Jackson, met and hit it off with Barksdale and the Governor, and got to work (Lewis 2006). In about three weeks, Duany and other CNU leaders built a team of 100-plus CNU members from around the country – including architects, planners, traffic and retail specialists – and brought them to meet an approximately equal number of Mississippi architects and planners at a more or less intact casino in Biloxi. In six days, between October 12 and 17, 2005, the Mississippi Renewal Forum – as it has come to be known – produced detailed proposals and plans for the redevelopment of the entire eighty-mile-long Mississippi stretch of the Gulf Coast devastated by the hurricane, divided into eleven communities between Waveland and Pascagoula. While obviously focused on a very different design and planning problem, the Mississippi Renewal Forum employed the same charrette methodology that DPZ had in designing the Kentlands in a process of the same duration. The CNU's ability to assemble such an extensive and carefully composed team, to work at such a large scale, and so intensively reflects the movement's development over the intervening seventeen years since the Kentlands charrette.

On October 18, the Forum's proposals were presented to area developers.[5] Final design team reports for each of the eleven communities were released a month later and two weeks after that "A Pattern Book for Gulf Coast Neighborhoods" came out (Mississippi Renewal Forum 2005). It contained examples of architectural styles that conformed to the traditions of the Gulf Coast and was made available for free at building supply stores and elsewhere to encourage rebuilding in a traditional vein. Some of the plans and renderings for proposed rebuilding were aimed at bringing attention and building opposition to the controversial new elevations above sea-level that FEMA introduced – showing old neighborhoods now on stilts, for instance, or proposing submersible houses. The FEMA rules required conformance to insure rebuilt private housing against similar disaster and both prevented rebuilding neighborhoods where they originally stood and interfered with the traditional spatial dynamics the New Urbanists were working to incorporate.

None of the work of the Forum was binding on the communities for which it was intended – that is to say, it was not, and is not, a given that the New Urbanist vision of the Mississippi Gulf Coast will be built – but it has shaped the debate about the process of rebuilding. In particular, New Urbanist involvement has focused local attention on the pre-hurricane "sprawling" of coastal towns and framed the rebuilding effort as an opportunity to undo – or at least re-consider – that tendency. Moreover, the planning process has brought New Urbanists into the sort of conflict with which they are familiar and in which they thrive – fighting established and rationalized convention in the FEMA rules. Finally, CNU's participation opened the possibility of a significant role for the New Urbanism in the rebuilding of coastal Louisiana and New Orleans, in particular. Various New Urbanist firms were recruited to begin recovery planning and design by private interests within the city of New Orleans and by municipalities on its borders. For the New Urbanism, Hurricanes Katrina and Rita may have been the "tipping point" (Gladwell 2000).

CNU's high-profile involvement in Gulf Coast rebuilding effort broadcast New Urbanist ambitions for greater influence and opened them to opposition on a new front. The Mississippi Renewal Forum web site has collected the press coverage of the charrette and subsequent events, and it encompasses hundreds of articles, most in local papers, but also from small and large papers across the country, including *The New York Times*, *The Washington Post*, *The Chicago Tribune*, and *The Los Angeles Times*. This is a potentially trajectory-changing development: the New Urbanism may now seem a known quantity to more of the country, a fact that can operate on the landscape in an unquestioned fashion.

Yet little has come of the Mississippi plans and less still of those for New Orleans. DPZ and Peter Calthorpe, among others, have provided input in New Orleans – run charrettes, developed plans – but on a smaller scale and with less publicity than in Mississippi. DPZ actually built four houses in the Bywater neighborhood of New Orleans as an example of what could be done – in a kind of competition with Brad Pitt's Make It Right Foundation which is building "150 affordable, green storm resistant homes for families living in the Lower 9th Ward when the storm hit" (MacCash 2009; makeitrightnola.org). The houses are thoroughly New Urbanist: styled to echo the neighborhood vernacular, but reoriented to capitalize on cross-ventilation for cooling and to produce semi-private space for their residents. They are also priced above what the local market will bear, apparently, and are being rented until the market improves (MacCash 2009).

Yet it is not clear that the failure to realize New Urbanist plans in Mississippi or New Orleans has at all damaged the movement's reputation. Their intervention earned them extensive publicity and reinforced the sense that the movement is "all the rage in town-planning circles" (Risen 2005). Nevertheless, failing to develop the New Urbanist solutions to the hurricanes' devastation will eliminate the ability of the movement to leverage built projects on this scale and in these circumstances, thereby by undermining the sort of synergistic progress discussed in Chapter 4. At the same time, of course, the failure to build draws attention to the political, economic, cultural and professional dynamics that interfere with New Urbanist development everywhere and potentially advances their position by providing evidence for their critique.

Another ambiguous contribution worth noting is the "Katrina Cottage." During the Mississippi Renewal Forum, a young New Urbanist architect, Marianne Cusato, designed a very small house to replace the ubiquitous and problematic FEMA trailer (Fornoff 2006; Lubell 2006; Stark 2006; FEMA 2008).[6] The original Katrina Cottage was 308 square feet, the same size as the trailers, and could be built on site – or ultimately assembled from pre-fabricated parts, when manufacturers were brought on board – for about $35,000, again similar to the FEMA trailers (Stark 2006). The intention was to provide better, cheaper, "context sensitive" housing" and was supported by a $400 million FEMA/Department of Homeland Security program (Lubell 2006). Nevertheless, while the cottages have gained currency, they were not bought in quantities by FEMA because they are "semi-permanent" structures, thereby violating FEMA's mandate to provide temporary emergency housing only. The hardware superstore, Lowe's, sold Katrina Cottage kits in the Gulf states, beginning in 2006, including substantially bigger versions of the cottage, but as of late 2011, they are no longer available (Lubell 2006; lowes.com, viewed November 28, 2011).

The Katrina Cottage was also the basis for the "Haitian Cabins," designed by DPZ for Haiti after the earthquake there in January, 2010, another disaster venue apparently ripe for New Urbanist intervention in new and different ways (Langdon 2010). Haiti presents some different possibilities for a number of reasons. The destruction was more complete than that in the Gulf, more major powers are involved in the question of reconstruction, and local government is more likely to be swayed by outside influence. In addition, New Urbanists are arguing for something that other experts have been championing for a long time, deconcentration (see Collier 2009; Lindsay 2010). Various commentators have arrived at the conclusion that a Haiti dominated by an over-dense, over-populated Port-au-Prince is less likely to succeed than one in which much of the population currently displaced by the earthquake is re-settled in other cities and towns or agricultural villages (Collier 2009; Langdon 2010; Lindsay 2010). Moreover, DPZ is fortuitously located in Miami, a center of the US Haitian diaspora.

It is too early to tell whether proposed New Urbanist interventions in Haiti will amount to anything. Early press coverage of DPZ's cooperation with housing manufacturer InnoVida to build and donate 1,000 Haitian Cabins does not indicate whether that has yet been accomplished – as of this writing. Their response to the Haitian earthquake should be understood as another opportunity for New Urbanists to build their public profile, but may, as we have seen in the case of the Gulf Coast, prove to be more discursive than concrete.

Conclusion

The New Urbanism has drawn significant critical attention from both ends of the ideological spectrum in United States politics and from architects, because its principles and processes intersect with those of various ideological positions and the profession, but it has, through a combination of pragmatism and groundedness in real places, managed to escape politicization or the coherent opposition of that profession. Moreover, the movement has engaged policies and events on the national stage actively and with significant recognition, if also with mixed results. HOPE VI and New Urbanist responses to Katrina and the Haitian earthquake undoubtedly magnified their public profile, but also highlighted some of the obstacles to implementation that they continue to confront. They got good press, but they have not, for the most part, gotten built, betraying a perhaps critical weakness.

In Chapter 6, we will turn to New Urbanist places, returning to the Kentlands in particular. We have seen above how the connection between the movement and its products protected it, but these products are home to thousands of people. Who lives there? Why? How does the inhabitation of New Urbanist places reflect and reproduce existing social relations?

6 Making New Urbanist places

We have considered New Urbanist developments as models, as referents, as bases for the movement's reform efforts in the political, economic, cultural, and professional realms. Here is our chance to consider them as *places*. Because the New Urbanism is both movement and places, how the places come to be and what they are like is critical to our sense of the phenomenon as a whole. Placemaking is, at least, the inhabitation of concrete constructions by people who incorporate their materiality into daily life, making these spaces into social environments. Yet that inhabitation is a process continuous with the processes of design, planning, permitting, and construction. Each element of producing the built environment anticipates its inhabitation and the social use of the place reflects its spatial dynamics. Here we will investigate how a New Urbanist place got made, in terms both of how it was planned to be lived in and how the collective life within it responded to the design's intentions.

To do this, we will return to the Kentlands. The transition from envisioning the Kentlands to inhabiting it paralleled the effort involved in going from agreeing on the principles of suburban reform to consolidating a consistent influence on the built environment of the suburbs. It involved engagement with opportunities and constraints and found resolution in the specific details of the material environment. While the Kentlands is only one New Urbanist development, and an early one at that, the struggles to complete it and the process of settling it point to complexities inherent in the New Urbanist project.

Having seen how the Kentlands was built, we will examine one of the key elements shaping its inhabitation: the pursuit of the sense of community. Whether or not New Urbanist designs produce that sense, the fact that they are designed to produce it constitutes a distinctive appeal, an opportunity for potential homebuyers, one that that shapes the choices those homebuyers make and how they then enact their inhabitation of the place itself. At the same time, research about the sense of community in the Kentlands suggests a close connection with homeownership and we will see that it proves extremely difficult to disentangle community from private property and the maintenance of property value.

Finally, we need to situate the Kentlands as place in its context: does it simply reflect the socio-economic and demographic trends in the region of the country in which it was built? Demographic data suggests the answer is no, it does not. The data are inadequate to rigorous causal claims, but they indicate that the Kentlands is quite distinct from its surroundings, leading us to speculate about why. A quick review of the scope and distribution of New Urbanist developments also reminds us that the Kentlands is one among many.

This exploration of placemaking in the Kentlands raises questions about the pursuit of a sense of community and adds force to the arguments of scholars more skeptical about community than the New Urbanists. Thus, in the conclusion, I explore David Harvey's argument that the pursuit of community may not be an unmitigated good, and we reflect on the ways that some of the Kentlands distinctiveness may arise from the concept's complexity.

Completing the Kentlands

We left the Kentlands at the moment when the process of envisioning and permitting it seemed to have succeeded. We return to it partly built, partly inhabited, and confronting challenges to its realization as designed. Changes in the larger political-economic environment in the country and the region required some adaptation of the developer's agreements with the City of Gaithersburg.

This section will show how the Kentlands project was completed and inhabited, but also how that process involved the active reproduction of a New Urbanist discourse of community through design. Moreover, it will locate the interplay of opportunities for and constraints upon the various actors in the concrete designs of the place itself.

NIMBY, changing ownership, and citizen action

Many Gaithersburg Planning Commission hearings that included consideration of the Kentlands project proceeded without any comment from the public, but the final meeting in early 1991 about an amendment to the Sketch Plan was different. Due to a significant downturn in the market for office space, developer Joe Alfandre came under pressure from his lenders not to build some of the office space he had planned and received approval for in the Kentlands town center. His lenders wanted him to build more residential units instead, in order to maintain the anticipated return on investment. The amendment, under consideration, would allow Alfandre to trade 1,000 square feet of planned office space for one residential unit, up to a proposed limit of 2,400 residential units, approximately an additional 1,000 units above what had been approved in concept. The application of this formula would be at the developer's discretion, but the new mix of uses would still be subject to city review at a second stage – the Schematic Development Plan review.

The confusion at this first hearing in which the Sketch Plan amendment was introduced triggered a kind of naked self-interest among both the laypeople on the Planning Commission and local residents that the project had mostly managed to otherwise avoid coming into conflict with. Beverly Weidmann, a Gaithersburg resident:

> ... expressed concerns about the proposed change ... due to the impact an apartment development would have on the adjoining residential community ... where she resides in terms of increased traffic, building mass, decreased property values and quality of living. She added she favors the originally proposed commercial mall on the northeastern portion of Kentlands.
>
> Planning Commission 1991b: 3

Martin McDonough – also a resident of Gaithersburg – it is reported:

> ... shared Mrs. Weidmann's concerns, and added that the proposed conversion ... indicated that the type of dwelling would be apartments, which he opposed due to transient residents. He requested that if changes to the Alfandre project are made, the dwelling units be single-family homes.
>
> Ibid.

No one spoke on behalf of the amendment, though the Planning Department did ultimately recommend it to the Commission.

Both comments reveal a common, NIMBY reaction to new development: a concern for property values and the potential introduction of an undesirable element. The amendment, in fact, did not specify that apartments would be built, specifying just residential buildings – townhouses and single-family homes included.

After these comments, the public hearing record on the amendment was closed – perhaps explaining why more public comments were not part of the official record. After some further discussion and Planning Director Russel's firm recommendation, the Commission agreed to recommend the adoption of the amendment to the City Council, who proceeded to adopt it at an April meeting.

What is remarkable is that this sort of NIMBYist, community and property-value defending public commentary is not prevalent *throughout* the official record of planning the Kentlands. American planners suggest it is the norm. Perhaps the difference is that public input seems to have had little influence on the initial phases of the process, except in the sense that the public was represented by elected and appointed officials. Nevertheless, Mrs Weidmann's and Mr McDonough's comments indicate the constraints within which Alfandre and his champions in Gaithersburg government were to come to work. Any suggestion that the project might depart from its apparent reinforcement of Gaithersburg's generally high-status housing market, even if only reflexively inferred from an increase in residential units, inspired a quick response. At the same time, the most important public involvement came after the plans for the project were complete and well under way, as described below.

Shortly after the Amendment passed, giving Alfandre needed flexibility from the city in the form of the office-to-residential exchange formula, he was forced to give the Kentlands project to the bank. While the work of building both infrastructure and housing in the first neighborhoods was proceeding, as were sales, the financial problems stemmed from the lack of an income stream necessary for maintaining his loans. The critical mistake would prove to be Alfandre's management of the mall component of the project's original design.

As we saw in our discussion of the Kentlands charrette in Chapter 3, Alfandre had sold part of the Kentlands property to a mall developer for more than 40 per cent of the total price he had paid for the land. Although the designers worked hard to make the anticipated mall fit with the greater vision for the project, the mall developer, Melvin Simon & Associates, walked away from the one million dollar deposit it had made toward the purchase of the parcel in 1991, leaving Alfandre with a much heavier financial burden than he had anticipated.

The problem was no one could find anchors – big department stores – who would commit to the site and make that aspect of the project economically feasible, despite the Kentlands' location at the intersection of two highways, a location that should have made it a plausible retail location according to standard real estate assessment methods. This failure to make the mall work, in the end, caused the financial downfall of the project. Bill Lennertz, a DPZ-employee at the time of Kentlands' design and development, explained: "the basis of the land was valued on a million square foot shopping mall. How the hell are you going to work your way out of that one? I don't think he [Alfandre] knew and finally it didn't work out" (Interview with author, May 23, 2007 [hereafter "Lennertz"]). According to Lennertz, the difficulty of integrating commercial and residential was a recurring problem for developers of mixed-use projects and Alfandre's inexperience as a mix-used developer was critical:

> the weakness shows in an area where the developer doesn't have experience and [Alfandre] hadn't done retail. So he didn't know retail, so he was held hostage by the retail community. … the real strength and the weakness of the project was that Joe was a residential developer.
>
> Lennertz

After "extended negotiations," in October 1991, Alfandre executed "a deed in lieu of foreclosure to Chevy Chase [Bank's] newly created, wholly owned subsidiary, Great Seneca Development Corporation [GSDC]" (Arkin 2004: 109). Russel, the Gaithersburg Planning Director, said, "And of course when he lost the property we were all devastated" (Interview with author, June 21, 2004 [hereafter "Russel"]).

Despite this major setback Kentlands became a New Urbanist place. The critical factor was that by this point the City of Gaithersburg's intimate involvement in the long development process had been transformed into a firm commitment to the project's neo-traditional principles. The city's Mayor Ed Bohrer was key. Russel continued:

> … after Joe lost the property in September of '91, well, [Mayor] Ed Bohrer made it real clear to the bank that unless they continued with the New Urbanist vision they weren't going to get any approvals. So they basically stuck with it.

Meanwhile, the bank found itself in an awkward position. The development process was far enough along that completing the project offered the best opportunity to salvage most of the bank's investment. At the same time, because development was under way and the plan was so distinctive, the bank was under great pressure from local government, residents, and homebuyers under contract to complete the project the way it had been designed and approved, despite the fact that elements of the concept, if not the design, had likely contributed to the project's financial failure. The bank's priorities initially conflicted with the New Urbanist design of the Kentlands, but it was ultimately the depth of homebuyer commitment that the design generated that enabled the bank to complete the project.

The bank's initial discomfort with the project as planned was immediately evident to their counsel, Jody Kline – who had worked previously for Alfandre. GSDC hired Kline to assist them in completing the development. He noted that working with them was quite different than his experience with Alfandre and that the bank's representatives were much more confrontational. Where Alfandre would wheedle and charm:

> [GSDC executives] Jeff [Campbell] and Page [Lansdale]'d call me up and say "Goddamn it, they can't do that. How can we sue these bastards?" And we never did. I mean, they learned that eventually it would all work out. It took them a while to get out of there though. In fact, they just sold their last lot there the other day [in 2004].
>
> Interview with author, May 13, 2004 [hereafter "Kline"]

More importantly, when the bank took over, it took a more cost-conscious approach, according to Kline:

> Chevy Chase [owner of GSDC], maybe with some basis in experience [since it had to take over the project when Alfandre defaulted], was saying, "Hey Joe, it didn't work, so we're not going to put ourselves in that jeopardy." So Chevy Chase was much more practical ... more pragmatic, in terms of what they did ... they were a financial institution ... trying to get out of basically a bad situation. So they were tighter.

This shift to an emphasis on the bottom line caused, as one would expect, tension with the city. Russel said:

> ... they were always trying to save money and we were always trying to do something a little bit better. We didn't feel like we should be compromised. And there were tons of compromises, we never had any granite curb. Just because Joe lost the property.

In an effort to address the tensions with the city, early residents, and homebuyers under contract, one of GSDC's early moves was to propose building the Kentlands recreation center. According to Russel, this proved a clever move in fostering allegiances with current and prospective Kentlands residents and galvanizing the support of the planning professionals who advised the Gaithersburg

Planning Commission, because the center brought together anti-conventional elements of the design and suggested GSDC's commitment to New Urbanism. The negotiations over the recreation center would smooth the way for future planning permissions necessary to the development because they would re-establish the New Urbanist foundations of the project after the change of ownership and a hiatus in building due to the transition.

At the same time, the permitting process for the proposed recreation center also shows how the New Urbanist vision was nearly undone by the conventions of suburban development as policed by the Gaithersburg Planning Commission, a body made up of elected laypeople tasked with enforcing community standards, not planning professionals. Once again, it was the buy-in of early Kentlands residents – buy-in established through the early charrette planning process that ensured the New Urbanist elements of the Kentlands project survived.

Russel remembered that when the Planning Commission was presented GSDC's plans for the recreation center, it responded badly. The problem was the unconventional lack of parking spaces necessary to the design of the center and its integration into the neighborhood: "And the Commission sort of went nuts and they went 'Oh God, you can't have a big facility like this with only sixteen parking spaces!' So, it's kind of like they forgot about New Urbanism briefly" (Russel).

Early Kentlands residents were key. Russel continued:

> And we enlisted the help of the then-residents, of which there were some. And it was great, they came that night to the hearing and they said, "We don't want more parking! We want it to stay green! We want the facilities! We want to walk! We want to bring our baby strollers! There's enough parking for us." And the Commission went, "Oh yeah, we forgot about that. This is a New Urbanist community." But it was the community working for itself, you know [and, in this case, for the bank].

This also highlights the multiple communities at work in the sense that it shows the Kentlands homebuyers in direct conflict with the community representatives of the Planning Commission. Requiring parking spaces, they argued, would increase the cost of the project, compromise the New Urbanist design, and reflect an unthinking application of conventional zoning regulations.

After the struggle over the recreation center was resolved, residential building slowly picked up in the development (Arkin 2004). A new shopping center developer, the Beatty Companies, was brought in to build the commercial part of the project, which, after significant discussion with the city and residents, they began. Wal-Mart and a drive-thru MacDonald's were prevented, then commercial development slowed down until it could be re-conceptualized as part of the Lakelands addition to the project in the mid to late-1990s (ibid.). By 2000, significant proportions of the Kentlands and Lakelands housing were occupied.

The success of the recreation center's approval points to the force of the community concept for the Kentlands residents, in that it overwhelmed a more diffuse Gaithersburg community conception focused on consistency in land use that is caught up with exclusivity.

The need for adaptation raised questions about the broader community's understanding of and commitment to New Urbanist principles.

In the face of financial difficulties, the pursuit of community was critical to the progress of the Kentlands. Even the bank, with its more singular focus on profit, efficiency, and the bottom line, could only realize its goals by navigating a complex course among real obstacles, in particular the enduring vision in which the city government and residents of the development had enrolled. This vision integrally involved New Urbanist "design for community" principles. Initially, the bank confronted the community concept as an obstacle in the city's inflexible demands about the completion of the development, but ultimately they capitalized upon it by leveraging New Urbanist design principle in the approval of the recreation center and thereby securing the support of the planners and residents. More, they built support by literally building an element of the project, resolving their engagement with the various opportunities and constraints through the specific materiality of the built environment.

The Kentlands story points to the power of community, but also to how community can operate at multiple scales and among multiple groups simultaneously. Community, in regard to the Kentlands, as we have seen, became a territory to be defended for some Gaithersburg residents, a design principle to be enforced for local planners, and a promised sense for potential inhabitants. It functions as an opportunity to be pursued by homebuyers, but, through their action in combination with the city, as a constraint on the developers.

The structure of community involvement

A key question that remains is: what made early Kentlands residents, in particular, such committed and effective advocates of the New Urbanist vision? The charrette process discussed in Chapters 3 and 4 certainly played a role. But the "community involvement," in the Kentlands as elsewhere, turns out to be neither neutral nor organic. Both the New Urbanist vision and the role of residents in policing it were thoroughly structured by the homeowners' association rules that homebuyers agree to when they buy. "Covenants, Conditions, and Restrictions" or CC&Rs are attached to the deeds of individual lots and mandate the participation of whoever owns that lot in the association, including paying fees.

Homeowners' associations (HOAs) are designed to protect the developer's investment by granting the developer substantial control over the project and its residents until the development reaches a particular stage of build out – usually when the developer's profits have been secured – at which point they shift legal and financial liability to the homeowners. Although they vary widely in how they do this, HOAs are frequently accused of being undemocratic, and are minimally regulated (for more on HOAs, see McKenzie 1994). The Kentlands homeowners' association structure played a key role in enabling residents to enforce a New Urbanist vision.

The Kentlands community's ability to work for itself effectively was rooted in a "very sophisticated" homeowners' association structure according to Duany, crafted by David Wolfe, the marketer and specialist in crafting homeowners' association documentation who introduced Duany and Alfandre. For one, it was unusually democratic. This empowered residents against the bank, when it took over. In Duany's words:

> The way they were written, it was very democratic very early on [as opposed to retaining power for the developer like in Celebration (see Frantz and Collins 1999 and Ross 2000)]. That was a big thing, because when the bank took over they weren't interested in keeping the quality up, it was just a bunch of bureaucrats. But the association documents were virtually unique in America in that they thoroughly empowered the citizens. [They] were the ones that upheld the quality: they bedeviled the builder and the builders the bank because they were highly empowered by this freak document. That was amazing, Kentlands couldn't have been the way it was.
>
> Interview with author, May 15, 2007 [hereafter "Duany"]

We can only speculate that Alfandre asked for this documentation because of his idealism about the project.

Although he does not emphasize the private legal structures specifically, Richard Arkin, who was one of the first Kentlands residents and one of the most involved, agrees with Duany about the citizens' role. In his self-published Kentlands history, he emphasizes the role of the community in the development process, specifically deemphasizing the roles of the Mayor and Gaithersburg planners:

> [By 1991 when Alfandre lost the project,] the dozens of homeowners and home purchasers already committed to the neotraditional 'Kentlands Vision' became a powerful lobbying force keeping city officials focused on completion of the project essentially according to plan. Kentlands residents became familiar faces at City Hall, poking through the files, speaking up at meetings, and lobbying individual staff members, commissioners, and councilmembers.
>
> Arkin 2004: 109

Arkin also suggested that the residents understood themselves foremost as monitoring the bank's activities. Buried in his account, however, is the key point: there were legally binding plans, which included design standards, to which to hold the bank. Thus, Arkin writes:

> I guess because I'm a bit of a utopian, I stepped forward to try to hold the Chevy Chase Bank people to the plan … . Most of Kentlands had been subdivided at that point. It was going to come off pretty much like it did, maybe it would have looked more like Lakelands if we hadn't hewed so closely to the design standards, but it would have been a success. Making sure people hewed to the design standards made it even more of a success, really an extraordinary artistic success. Which I think reflects itself in the property values here too.
>
> Interview with author, June 19, 2004 [henceforth "Arkin"]

Thus residents were empowered to get involved in policing aspects of design. As Russel recounts:

> We had some very unattractive evenings ... there were times when the contract purchasers [homebuyers whose houses were not yet built, but under contract with a builder] would be in the audience and the commission or one of the residents of Kentlands would be there policing the aesthetics. And the commission wouldn't approve it or the people would be crying or they'd be negotiating in the background with their builder, because the commission wanted more brick and they didn't want to pay for it. So there was a lot of that kind of stuff going on, because the people who moved there felt very strongly about it and they moved there for a purpose.
>
> Russel

The homeowners' association structure enabled a community conception that facilitated the involvement of Kentlands residents in pressing GSDC, the city government, planners, and the Planning Commission for the completion of the project along the lines on which it was begun. This included holding the design of houses still under construction to a higher standard than the bank necessarily would have, because even houses under contract remain subject to the approval of the design committee of the Kentlands Citizens Assembly, per the CC&Rs. As Russel's comment implies, of course, that quality here is synonymous with cost, a theme we will return to below. In sum, the profound degree of enrollment in the vision of Kentlands shared by homebuyers and the city government was critical to the development of the Kentlands as a New Urbanist place despite significant pressure to abandon the vision when it proved an economically questionable endeavor.

Sense of community in the Kentlands

New Urbanist developments like the Kentlands advertise their community-building intentions loudly and claim that distinctive characteristics of the designs and plans produce a community effect. This is their market advantage. Unsurprisingly, researchers have attempted to test these claims. Conveniently, important entries in this research have been done in the Kentlands itself.

Architecture scholar Joongsub Kim and social anthropologist Bengt Andersen separately completed extensive observation, interview, and survey-based research on which elements of the Kentlands design and plan produce a sense of community among residents and about the sense of community more generally (Kim 2000, 2007; Kim and Kaplan 2004; Andersen 2001; see also Nasar 2003). Kim's evaluative research, in particular, is thorough and thoughtful, but suffers from a couple of fundamental methodological problems. First, the residents of New Urbanist developments are overwhelmingly aware that the design is supposed to produce a sense of community in them, because they were told it would when they moved in. This makes separating the design's actual effect from the residents' predisposition to report it quite difficult. Second, the administration of survey instruments, no matter how thoughtfully designed (see Kim 2007, for example), could itself facilitate the connection in the minds of those surveyed between New Urbanist designs and sense of community simply by asking about that possibility.

That said, Kim's findings do suggest that the sense of community is importantly bound up with homeownership itself. Kim's (2007) central conclusion, from extensive survey and interview data gathered in the Kentlands in 2000, is that a "sense of ownership of home and community" is strongest among those in the development who are most likely to own their homes, by comparison to renters. Andersen (2001) finds similarly that the New Urbanist ideology and sense of community are tied up with the practice of protecting property values in the Kentlands (see Arkin interview below). Though Kim is at pains to point out that a "sense of ownership of home" is not identical to the legal and financial fact, it is clear they are closely connected in the Kentlands case.

These findings are hardly surprising. The pursuit of a sense of community is a distinctive element of buying a home in a New Urbanist development, as opposed to in a conventional one, because is it so clearly articulated as a designed-in aspect of the project. As homebuyers engage in the process of imagining home, envisioning themselves and their families in these new surroundings, they can construe the benefits of the sense of community they will purportedly experience. Moreover, because the New Urbanists define the concept of community broadly, if not purposely vaguely, potential homebuyers can imagine community in their own image. Put differently, the vagueness facilitates New Urbanist homebuyers' self-selection for that sense, because people can define community for themselves, and makes it that much more likely they will experience their version of it once they live there. For the broader purposes of understanding how New Urbanist places get made, testing whether or not the design produces a sense of community seems less relevant than examining how homebuyers engage the sense of community as a goal or an opportunity, because the process of self-selection for a sense of community will prove one of the primary sorting mechanisms among potential New Urbanist homebuyers.

My informants clearly indicate the degree to which Kentlands residents self-select for the pursuit of the sense of community. Mike Watkins, DPZ architect and Kentlands resident, said, "Very early on, the community attract[ed] people who value[d] community, not surprisingly, so these people moved in ready to engage themselves in the lives of their neighbors, in their community and in the city" (Interview with author, December 5, 2005 [hereafter, "Watkins"]). Judy Howlin, realtor and Kentlands resident, commenting on Kentlands homebuyers, agreed, "If there's a common thread of people here, it's people that like interactivity. You don't find very many loners or hermit-type people living here because they're people that like to see other people and see the activity going on" (Interview with author, June 22, 2004 [hereafter, "Howlin"]).

Both interviewees emphasized throughout our conversations that the physical arrangement of the development was critical to the reproduction of this tendency among current residents. People moved in ready to "do community" but they interacted with one another a lot because of the proximity to one another in which they lived, among other reasons.

Community involvement, to be sure, does not always prove harmonious. Richard Arkin, Kentlands resident, former Citizens Assembly president, and Kentlands historian, reminded me that interactivity and the pursuit of community can cut two ways, "There are lots of opportunities for friction as well as friendship. … Less privacy here [than in conventional suburban subdivisions], everybody knows everybody's business and there are more opportunities for people to clash" (Arkin). My other informants suggested obliquely that Arkin was someone with whom others had frequently clashed.

More interestingly, informants offer further evidence of the complicated relationship of community engagement with homeownership. Both Arkin and Watkins pointed out that some of this interactivity –both cohesive and clashing – is focused on citizen maintenance of design standards. While design standards are integral to New Urbanist principles, they are also sometimes proxy for, or at least entangled with, property value. Watkins remembered:

> In fact, I have to say that [while] some of the residents took the time to inform themselves, [to learn] what was important about the traditional neighborhood, others came at it with their own agenda, which was really *not* to see a mix of housing types, but to see houses that sold for as much or more than theirs did, so that their property values weren't threatened. And there was some misunderstanding about which battles to choose.
>
> Watkins

Arkin said, with a different emphasis:

> Well, property value is never far from peoples' minds. Property values have increased more sharply in Kentlands than in surrounding neighborhoods. … And I think in part that's because of the physical attractiveness of the community, and I think what motivates those who have been on the two design boards has been to maintain the physical attractiveness. … It's real hard, when it comes to property values, to pin down what it is that makes a community a desirable place.
>
> Arkin

Watkins, speaking as a New Urban designer, sees the pursuit of property value interfering with the realization of New Urbanist concepts like the integration of housing types and sizes, while Arkin, speaking as a homeowner, makes clear that enacting community is often synonymous with the maintenance of property value. This is not only that the sense of community and property ownership are intertwined, but that having self-selected for the pursuit of a sense of community, Kentlands homebuyers frame activities as community building that conventional homebuyers might not identify the same way.

All of my informants, moreover, suggested that the pursuit of a sense of community clearly distinguished Kentlands homebuyers from those who settle in conventional developments. About early residents, Watkins said:

> I can't think of a single person who shared the principles that we proposed and that somehow they fell out ... who misunderstood what they were getting into. ... You don't buy a house on a lot the size of those in Kentlands by mistake. You know that you're giving up square footage for a neighborhood.

Howlin said similarly, about potential homebuyers who tour the Kentlands and refuse to consider buying there:

> They don't get the concept of what's trying to be created. ... And sometimes they'll even say it in a derogatory way: "Why do people want to live so close together? Why do they want houses on top of each other? This house is $700,000? I can get a much bigger house for 650 and it's got two acres of land!" And I'll say, "Go ahead."

And Arkin summed up, "A lot of people who live in conventional suburbs ... they hate this place". Not only do Kentlands homebuyers make the pursuit of a sense of community a priority, but they are making sacrifices to undertake that pursuit, at least in some conventional assessment of real estate value, which of course exerts further pressure on them to realize it.

To summarize, the pursuit of the sense of community is an integral part of the process of buying into the Kentlands. The perception of the sense of community, *pace* Kim and Andersen, is bound up with the sense of ownership and the activities of ownership – in a for-profit real estate market. And all of this is further motivated by the awareness of economic sacrifice necessary to buy in the first place. If we can't disentangle sense of community in the Kentlands from the financial and the legal, perhaps it is as much a political-economic phenomenon as an interactional or social-psychological one. Kentlands may be an extreme case, because of its early development or for reasons related to its location, but it should be understood as indicative nevertheless.

The Kentlands in context: scope, distribution, and difference

Scope and distribution

The Kentlands was the most important early New Urbanist place, because, unlike Seaside, it showed that New Urbanism worked in a typical suburban location and, unlike Celebration, it could be developed like a conventional project, not requiring a gigantic institution behind it. Tours of the Kentlands, as we have seen, served a critical function in spreading the New Urbanist vision among planners, developers, and municipal officials.

Even as the Kentlands is prototypical of New Urbanist, neighborhood-scale developments, though, it is only one among many. *New Urban News* began tracking the number of "neighborhood-scale" New Urbanist projects in 1995. The editor, Robert Steuteville, defines "neighborhood-scale" as "must cover at least 15 acres, feature an interconnected network of streets and a mixture of housing types, and be organized around at least one central gathering place"

(Steuteville 2005). In 2002, projects tallied ranged from 15 to 12,400 acres, averaging approximately 375 acres. As of the January/February issue of 2005, *New Urban News* reported its inability to continue providing an annual tally because of the rapid proliferation of projects, combined with the difficulty of evaluating them qualitatively. That issue reports 760 projects, but calls the number incomplete and fails to break it down as had been done previously. More recent estimates suggest more than 1,000 projects.

Three issues of *New Urban News* also report breakdowns of the total projects tally by region of the country. This makes very clear that the New Urbanism has been a regional phenomenon, by far the most successful in the Southeast, followed by the Mid-Atlantic, Pacific, and Midwest.

In the years from 1995 to 2003, the change in the number of projects built or under construction as a percentage of those planned in the previous year increases steadily from under 20 per cent to nearly 50 per cent. This suggests that success breeds success – that moving projects from the planning stage to the construction stage makes repeating that step all the more likely in the future. This suggestion is consistent with my other findings about the relationship of the New Urbanism's progress to built places.

The role of sucesss in breeding more success likely explains the regional effect as well. Many developers are organized regionally, whether small-scale local operations or sub-divisions of national or international ones. Success in particular regions is thus likely to amplify continuing success in the same.

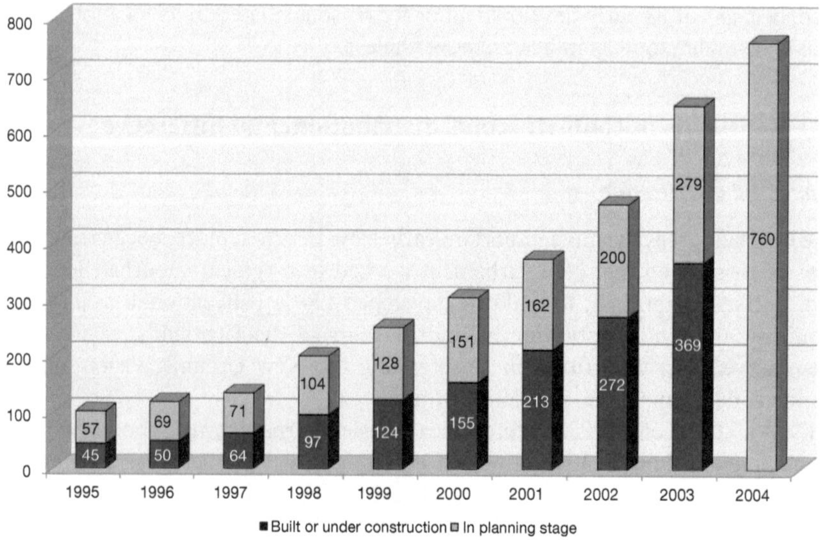

Figure 6.1 New Urbanist projects 1995–2004

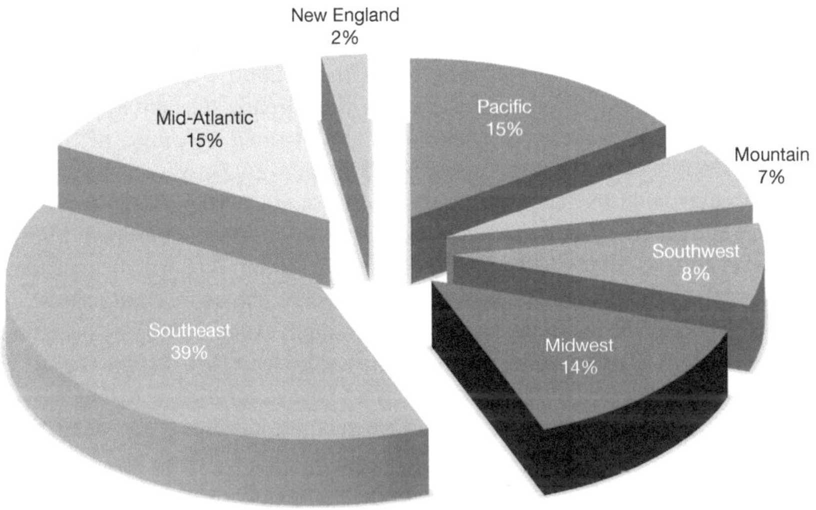

Figure 6.2 New Urbanist projects by region, average 2001–2003

The Kentlands difference

Just as the making of the Kentlands reinforced the New Urbanist movement and its proliferation, so too do the movement's principles and plans shape the process of placemaking in the Kentlands itself. That is, the New Urbanist design of the Kentlands also *did* things for its residents. One way to provisionally answer what it did is to locate the Kentlands in the broader demographic context of the United States and its surrounding region. When we do so, we find that what the Kentlands seems to do is exaggerate its distinction, exaggerate its difference from the rest of the country, even from the relatively wealthy county within which it is located.

Despite or because of the New Urbanists' best intentions, the Kentlands appears to consolidate privilege and homogeneity. Demographic data show that the inhabitants of the Kentlands are Whiter, wealthier, more educated, and higher-status workers than the average American, but also than the average Marylander and even the average Montgomery County resident. They also live in more expensive houses than the average American, Marylander, and Montgomery County resident. More than that, as the United States, Maryland, and even Montgomery County have become more diverse, the Kentlands and Lakelands seem to stabilize and consolidate their relative distinction.

This trend is confirmed by the analysis of demographic change over time. I conducted a nested comparison of data from the 1990 and 2000 Censuses and the 2005–09 American Community Survey, and race and age data from the 2010 Census – available through Social Explorer. I compiled data on racial composition, median household, family, and per capita income, educational attainment, occupation, and median home price for the United States, Maryland, Montgomery County, and the census tract or tracts that includes the Kentlands and the Lakelands. The findings of my comparison are consistent across time period and across indicator.

The greatest changes appear between 1990 and 2000 – before the Kentlands was fully established as a place, but when it was already substantially complete. While the United States, Maryland, and Montgomery County all declined consistently in their percentage White by 10–20 per cent between 1990 and 2010, the Kentlands and Lakelands tracts increased from over 70 per cent to nearly 80 per cent White. In 1990, there were more Asian-Americans present at both the tract level and the county level than in the state or nation. By 2010, the percentage Asian-American declined at the tract level to below the county proportion. The same trend holds for the Black population although, even in 2010, the percentage Black remained at approximately 4 per cent in the tract, low for the region, but not entirely absent, despite financial barriers to access discussed below.

When we look at a narrow band of high earning families – classified by the census in three groups $100,000–149,999, $150,000–199,999, and more than $200,000 – we see that families with an income greater than $100,000 increased in the United States and Montgomery County from 5.4 per cent and 16.4 per cent in 1990 respectively to 26.4 per cent and 56.2 per cent in 2005–09. In the Kentlands and Lakelands tracts, however, they increased from 16.2 per cent in 1990 – prior to the construction of the Kentlands – to 73.4 per cent in 2005–09. Put simply, prior to the construction of the Kentlands the census tract had marginally fewer high earning families than Montgomery County, with 16.2 per cent compared to 16.4 percent, whereas by 15–19 years later, the Kentlands and Lakelands tracts have a significantly greater concentration of high income families than the rest of Montgomery County (73.4 percent compared to 56.2 per cent).

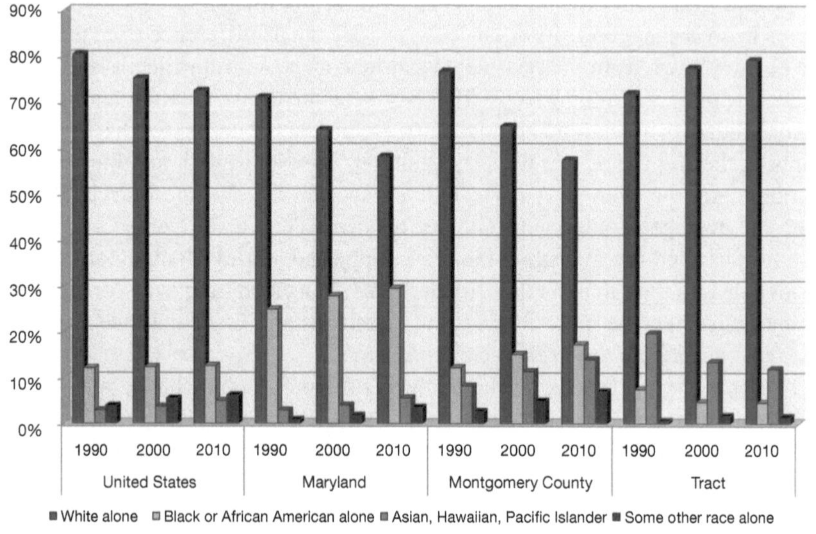

Figure 6.3 Racial composition

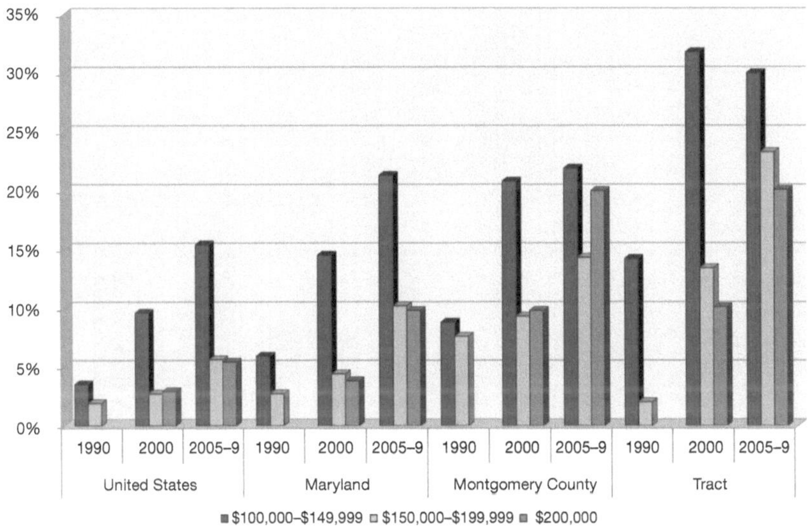

Figure 6.4 Family income

A similar trend holds when the measure of comparison is property values rather than income. Median home value more than doubled across the United States in this period, even taking into account some effect of the housing downturn, but in the Kentlands and Lakelands the increase was still more drastic. This holds true even if we restrict our comparison to Montgomery County, which is a wealthy county and, thus, by no means representative of the United States. In Montgomery County, median home value increased from $199,000 in 1990 to $487,500 in 2005–09, while it more than tripled in the Kentlands and Lakelands tracts, from $181,500 to $562,800.

Perhaps not entirely coincidentally, the ratio of highly educated people in the United States to those in the Kentlands and Lakelands behaved similarly. While the percentage of the population with a BA or higher degree increased from 20 per cent in 1990 to 27.9 per cent in 2005–09 in the US as a whole, it increased from 60.9 per cent to 69.8 per cent in the Kentlands and Lakelands in the same period. Between 2000 and 2005–09, i.e., after the Kentlands was established, there is clear indication of stabilization on this measure.

A precise causal story is hard to tell here and the data considered above only enable us to speculate. Nevertheless, given their consistency across indicators, they strongly suggest that the Kentlands and Lakelands have facilitated the self-segregation of a high status population. Rather than integrating different kinds of people through increased casual interaction in their neighborhood, the Kentlands and Lakelands appear to enable those who are financially capable to live among others like them and to pursue a sense of community among those like them. Whatever there is to admire about New Urbanist, suburban, greenfield

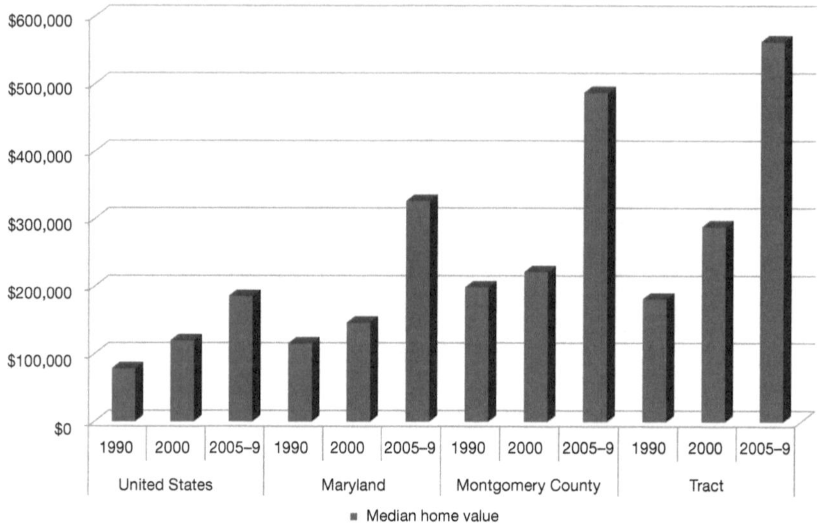

Figure 6.5 Median home value

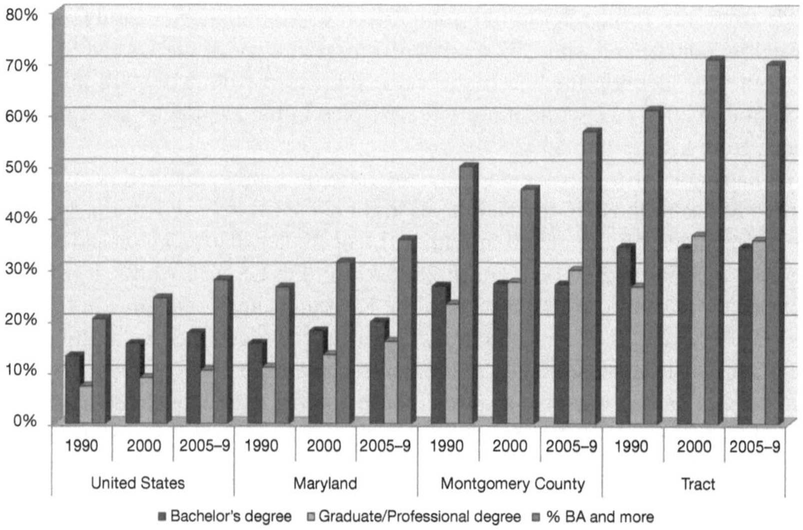

Figure 6.6 Educational attainment

development, its has not succeeded in undoing, or even mitigating, the socioeconomic and ethnic segregation of the metropolitan region.

Housing prices do much of the work here, illustrating the ways in which whatever opportunities the Kentlands specifically – or New Urbanism more broadly – might offer in terms of changing American suburbs are constrained by entrenched structural constraints.

Higher home values disproportionately exclude people of color from buying into a New Urbanist place like the Kentlands. The ability to afford a home is based in income and wealth, and gaps in both between Whites and people of color, particularly Blacks and Latinos, are substantial and have been widening in recent years. It should be noted that the Kentlands and Lakelands include rental properties so the barriers to local residence are not identical to those to homeownership. Still, homeownership is a key part of the picture.

The US Bureau of the Census reported that in 2010 median household income for Blacks ($32,068) was 58.7 per cent of that of non-Hispanic Whites ($54,620), a slightly increased gap over that in 2009. This is largely because Black income declined more significantly during the Great Recession than did White income, although Black median household income has been relatively stagnant at roughly 60 per cent of White median household income for more than forty years (US Census 2011, as cited in JBHE 2011). Emily Talen confirms this observation in recent research that incorporates data from 153 New Urbanist developments (Talen 2010).

Income is only part of the picture. More starkly, the wealth gaps between Whites and Blacks and Whites and Latinos in 2009 were 20-to–1 and 18-to–1 respectively (Taylor et al. 2011). These differences are substantially explained by the fact that homeownership rates are significantly lower among Blacks (46 per cent) and Latinos (47 per cent) than among Whites (74 per cent) in 2009 (ibid.). Moreover, these 2009 ratios are almost twice the gaps that prevailed for the twenty years prior to 2009, which were themselves obviously substantial, and represent primarily a radical loss of housing wealth for Blacks and Latinos associated, once again, with the Great Recession (ibid.). The key here is that the greatest single component of household wealth for most Americans is housing wealth. The wealth gap is, then, both caused by differences in housing wealth and contributes to the different abilities of racialized populations to build housing wealth.

Wealth and income gaps simply put relatively expensive developments like the Kentlands and Lakelands out of reach of most households of color. According to the real estate information service Zillow, the historical price-to-income ratio in the United States is just less than 3-to-1, though the ratio approached 5-to-1 prior to the downturn in 2007, returning to 3.31-to-1 in 2011 (Guddell 2011; Engebreth 2012). Calculating housing affordability as a function of household income is harder because mortgage rates vary across time and around the country, but based on the consensus maximum expense of 30 per cent of income on housing, affordability ranges from about 3-to-1 to about 5-to-1 (benengereth.com). This reiterates that $550,000 median home values in the Kentlands between 2005 and 2009 effectively preclude the in-movement of Blacks when only 1.1 per cent of Black households earn more than $200,000 (JBHE 2011).

The filtering effect of housing prices is likely reinforced by homebuyers' residential preferences for neighborhood composition. While 83 per cent of Blacks reported their preference for racially integrated neighborhoods in 2008 (Pew 2008, as cited in Bader 2012), recent research has shown sensitivity to a racial tipping point for neighborhoods and suggested that Blacks are less likely to select neighborhoods when the Black population declines below a significant proportion (Bader 2012). On the flip side, another recent study suggests that White families with young children are much less likely to settle in integrated neighborhoods, suggesting that, to the degree the New Urbanism is targeted at young families, it may again be exaggerating de facto residential segregation (Goyette, Iceland, and Weininger 2012). Consumer preferences, in the face of structural constraints, in other words, are likely to play an important role in explaining the trend.

Conclusion: community as constraint

The concept of community, while underspecified, is at the center of New Urbanist claims to reform suburbia and, if the Kentlands is any indication, it is central to New Urban homebuyers' engagement with their social and material context. Community as opportunity and the pursuit of it distinguish New Urban homebuyers and also enable them to distinguish themselves. Nevertheless, community also operates as a constraint: upon both the design and development process of New Urbanist projects, and upon the developers and funders of those projects, especially when residents and other locals are empowered to intervene in the development process.

More subtly, community operates as a constraint upon who will choose to live in New Urbanist places, upon who imagines themselves at home there, and upon how the inhabitation of New Urbanist projects is actually undertaken. This is, in significant part, because the sense of community is difficult or impossible to disentangle from property ownership and the maintenance of property value inherent in the broader, political-economic and cultural condition of the market for housing in the United States, reinforced by the omnipresence of HOAs. The intricately intertwined histories of politics, economics, and race in the US housing market mean that only some few – mostly White, upper middle-class, and well educated – are likely to be in the financial position to consider residence in the Kentlands and from them self-select the core of the new community.

Access to full participation in the Kentlands community, in other words, is constrained by the complex history of housing finance and this should remind us that the goal of community itself deserves some skepticism. The Kentlands is not representative of suburban, let alone urban, life in the United States, but the entangled-ness of community with political and economic factors in the Kentlands case is likely indicative of the complexity of the pursuit of the concept in most cases.

David Harvey's indictment of the New Urbanist quest for community (mentioned above in Chapter 5), is somewhat overstated, but the story of the Kentlands is a reminder that the concept sometimes fulfills a darker purpose than that for which the New Urbanists celebrate it. Harvey wrote, "'Community' has ever been one of the key sites of social control and surveillance, bordering on overt social repression. Well-founded communities often exclude, define themselves against others, erect[ing] all sorts of keep-out signs (if not tangible walls)" (Harvey 1997: 3). He suggests that the pursuit of community is a dialectic process – the inward turning search for connection is frequently paired with an impulse to distance oneself from the outside. Harvey continued, " … the New Urbanism … builds an image of community and a rhetoric of place-based civic pride and consciousness for those who do not need it, while abandoning those that do to their 'underclass' fate" (ibid.).

Even scholars more sympathetic to New Urbanist goals recognize this tension in the concept of community. For example, Robert Putnam notes that community frequently relies upon "bonding" social capital as opposed to "bridging." He explains this distinction as one between reinforcing exclusive social ties within similar groups – bonding – compared to one forging connections across networks and across social difference – bridging (Putnam 2000).

While the primary constraint upon homebuyers in regard to the Kentlands, and perhaps New Urbanist developments more generally, is cost, the concept of community is not without cultural and political resonance. Both Andrew Ross (2000) and Douglas Frantz and Catherine Collins (1999), in their studies of neotraditional Celebration, Florida, note without surprise that Disney's nostalgic Southern town failed to attract middle-class, Black homebuyers, to the apparent disappointment of its developers. Celebration is an extreme case, in that the marketing of its traditional design relies heavily on a harkening back to a "better time" that was not a better time for Americans of color. It is interesting, therefore, that the Kentlands and Lakelands, which did not partake in such rhetoric, have also failed to attract large non-White populations. Does "community" not sound equally appealing to all potential homebuyers?

New Urbanists propose to address the history of racial, ethnic, and class segregation in the metropolitan region by building places that foster community by design. There is no evidence to suggest that they are anything but honest in their idealism. Yet they are constrained by the for-profit housing market. Given this, their continued insistence on the ability of a sense of community to overcome many of the social ills that they identify in the context of the market begins to seem like a willful ignorance.

7 Conclusion

The built environment is a social process

The purpose of this book has been twofold. First, it investigates the emergence of a phenomenon that is fascinating in its own right, particularly from a sociological perspective. The New Urbanism claims to create community through design, embodies a significant shift in direction for suburban development in the United States, and represents a complex, collective effort across professional boundaries. Moreover, it makes places relatively appealing even to this elitist, urbanist eye. In doing all of these things, the New Urbanism reveals precisely how complex the social-material relationship is within the built environment. This is its sociological appeal.

Second, the book suggests a framework for analyzing the social-material relationship in general. An account of the built environment attentive to the opportunities and constraints imposed on its production by political, economic, cultural and professional forces and grounded in its explanation by the concrete details of the material instantiation of that environment accomplishes a number of fundamental things. It addresses the so-called "structure-agency problem" that persists in sociology by bringing together large-scale social forces and the activities of particular people, showing how individuals are key to the creation of movements and places but also how the activities of individuals reconfigure and reproduce social structures. It works the carefully formed, pre-conscious practices of various professionals into the social-scientific conversation about development that has been dominated by political-economic and cultural analyses, describing the important role played by professional agents and explaining its relationship to their professional formation. In doing so, the framework shows how the material is social by establishing the continuity between the complex process of its production and the equally complex process of its inhabitation.

The book does this by complicating our sense of how many social factors are involved in these processes. The vignettes in the first chapter set the stage by affirming the critical experiential quality of the built environment. We should never forget that our physical environments are incorporated into social life at the level of practice and that the characteristics they embody – political-economic, aesthetic, etc. – inform that practice as well as simply shaping it.

The second chapter lays out the relevant historical conditions out of which the New Urbanism emerged. While this is no doubt a simplified account, leaving out important historical events and trends, I assert that the elements of the historical context emphasized here were the most influential. I do so, moreover, to invite debate about these details. The central contribution of my account, as I see it, is identifying the importance of architects and planners. Unlike more conventional historical accounts, mine incorporates the dynamics at work within these professions involved in the design, planning, and construction of the built environment.

Chapter 3 reconstructs the origins of the New Urbanist movement through the accounts of the founders themselves. In other words, it does the "agency" part of the structuration process, showing people doing things. The things they did did not necessarily lead directly to the outcomes they intended to produce, but the combination of efforts described constituted the movement as it came to be. The diversity of explanations offered by the founders points to how different members of a collective effort remember and explain the course of historical events differently, but more importantly shows how the outcomes were as much a product of conflict and contestation as cooperation.

Chapter 4 brings structure, history, individual and collective action together to lay out the framework for the analysis of the production of the built environment discussed earlier. It argues that the New Urbanist movement emerged out of a particular historical context with a distinct set of political, economic, cultural and professional conditions, and that the New Urbanism confronted these as a set of opportunities and constraints. Actors within the movement engaged these opportunities and constraints and built the movement upon them, around them, and in response to them.

Critically, though, they did this by literally building. Making places provided an opportunity to resolve diverse tensions in the specific details of concrete designs, design practices, and the process of inhabitation. The New Urbanism was always a collective effort to reform the development process by doing development differently.

I play out this framework to explain the emergence of the New Urbanism, but I argue that it is generalizable to other cases. The framework owes its theoretical foundations to Bourdieu's concepts of field, practice, and habitus; Latour's translation model – enrollment in particular; and Molotch's lash-up, as discussed in detail in Chapter 1; and it applies these profound insights to urban and suburban studies in a way that has not been done previously.

The fifth chapter takes up the question of what happens to the New Urbanism once it has emerged – how does it survive? One important aspect of this question is how the New Urbanism has navigated pre-existing ideological conflict over city and suburb. I show how the movement escaped politicization through its comprehensiveness, but even more so, its pragmatism. I also considered how the movement weathered criticism from design professionals upon whose territory it has encroached and the recent emergence of alternative approaches for reforming urban and suburban development, notably Landscape Urbanism.

Chapter 6 reminds us that the movement is built around the creation of places and shows how placemaking involves a process of inhabitation. We returned to the Kentlands, unpacking the steps of its completion and exploring the role that New Urbanism's central selling point – the creation of a sense of community – played in the process of placemaking there. Rather than evaluating whether the New Urbanism does, in fact, create a sense of community, I show that the discourse around community was important in a range of ways, including how the residents of the Kentlands imagined their home and preserved its value.

Another key aspect of the social-material relationship is more suggested in Chapter 6 than demonstrated: the manner in which the built environment is caught up in the reproduction of existing social relations. It seems the Kentlands and Lakelands reinforce the privilege and distinction of their residents. This is not a novel observation, but the Kentlands case expands our understanding somewhat. Social historians of urban and suburban development like Hirsch in *Making the Second Ghetto* (1998) and Jackson in *Crabgrass Frontier* (1985) have shown us how the construction and inhabitation of the built environment – Chicago subsidized housing projects in Hirsch's case, post-war suburbs in Jackson's – is part of a process of racialization and the pathologization of poverty in the United States, keys to the enduring class, racial, and ethnic segregation of the metropolitan region. More recent critical studies of gentrification reflect a similar understanding (see Lees, Slater, and Wyly 2008, for example) as does work on recent ethnic transitions in inner-ring suburbs. My research suggests how the most banal of American environments – home, moreover, to the majority of Americans – reinforces axes of social difference. This is importantly related to the role of the private market in housing – and real estate in general – but we can speculate about what the concept of community, in the New Urbanist case, communicates to homebuyers from historically excluded groups.

Much of this study is caught up in questions of interests, both those clearly articulated and intentionally pursued and those that go unnoticed, unexpressed, or even remain unconscious. New Urbanists, design, planning, and development professionals, government officials, homebuyers, homeowners, and critics, at least, pursue their interests directly and indirectly, successfully and otherwise across the pages of this book. My informants' own agendas must inflect the insight they provide. All of which is a reminder of the difficulty of representing and analyzing a complex empirical world.

Questions that remain

A number of important questions are beyond the scope of this book. The intricacies of the Kentlands case point to the need for comparative studies across geography and recent history. Are the same processes at work in Prospect, a New Urbanist development in Longmont, Colorado? It is smaller than the Kentlands, but designed by DPZ and begun only about five years later. How about Stapleton in Denver – much bigger and slightly more recent (begun in 1998)? Do the challenges change as New Urbanist developments demonstrate their viability and as political, economic, cultural, and professional conditions change? Furthermore, the New Urbanism itself is obviously not a static entity. How has it continued to grow and adapt?

While case histories would enable the comparison of the development processes, quantitative comparisons could confirm or dispute my tentative conclusions about the Kentlands and Lakelands distinctiveness within local demographics.

Geographic Information Systems (GIS) software makes possible the comparison of a large number of New Urbanist projects, their relationships to one another, and to the regions in which they were developed. One could examine change over time in a large number of widely distributed places to assess the relevance of region, scale of development, location in relation to large cities, and demographic trends. This would also be a good way to approach the question of whether the Great Recession affected conventional and New Urbanist developments differently, a claim that follows from some New Urbanist assertions.

This study was also, of course, conspicuously American. Not only has the New Urbanism gained significant traction in Anglophone Canada – perhaps more comprehensively affecting the planning establishment there than it has yet in the U.S. – and in Australia, but it has an important ally in the Prince of Wales and cognate efforts in the U.K. In addition, suburban development in various forms, including some influenced by the New Urbanism, is taking off in Latin America, India, Turkey, and China, to name just a few. How does it compare?

This study raises questions too about other contemporary efforts to change cities and suburbs. I discussed the Landscape Urbanism in Chapter 5, but it likely deserves a more comprehensive treatment as it continues to constitute itself, whether it succeeds or fails. And what else? Are there other attempts to change how we make places distinctively affecting how we do so? And are they, in fact, susceptible to comprehensive portrayal through the framework I propose? How will the New Urbanism fit in or adapt as alternatives arise? Will it hold together as a movement or scatter as a number of widely embraced principles whose origins and interconnection are forgotten?

The Kentlands, 2012

I visited the roughly 20-year-old Kentlands – and 15-year-old Lakelands – on a sunny spring afternoon to take some photographs to illustrate this book. I ate lunch at a Persian restaurant in the shopping center, then walked the perimeter of both developments and crisscrossed through the middle. A number of things stood out.

Most importantly, it is clear that the Kentlands and Lakelands have become real places. As I walked around I kept thinking that I had only scratched the surface in my account. Even to talk about "the Kentlands" as if something existed independent of the many lives that overlap its geography is part of the problem. This is, all the same, the problem of sociology: we create abstractions to be able to analyze and theorize social phenomena, but in doing so something is lost. Ultimately, I felt both that shortcomings would be inherent in any account and that the attempt was nevertheless worth the effort, because the Kentlands and Lakelands fascinated me again as I walked around.

Next, the Kentlands and Lakelands are really done and, more, have been done for a while. Lakelands school and park, just red dirt the last time I was there, are complete and occupied and there's a new synagogue nearby that I hadn't known about. In the Kentlands, paint is chipping from weathered exteriors and porch floors have started to buckle from long-term exposure to the elements. As if to indicate the developments' maturity, the kids I saw walking home from school on this particular afternoon weren't the eight and nine-year-olds I'd encountered previously, but teenagers, if young ones.

Lastly, as if to confirm both the benefits of community and its drawbacks, it took less than two hours for me to be stopped by a police officer. He had received numerous calls about me as I walked briskly around the developments, stopping occasionally to take pictures and make notes. He mentioned my attention to the synagogue as a particular concern. I had anticipated this and did my best version of "White and polite:" apologizing, providing university ID without being asked, and explaining that I was taking photographs for my forthcoming book. Once past his initial need to ascertain my harmlessness, the officer explained that he just lived there "because it was a nice neighborhood" and marveled that people came from all over the country to visit the Kentlands and Lakelands, let alone write books about it. We parted with a friendly handshake, though he also warned me that if I kept it up, he'd be getting more calls.

Underlying my initial interest in the New Urbanism as a movement dedicated to creating new places and building community is an old urbanite's disbelief in the possibility of authentic new places. ("Authentic" is, of course, a concept perhaps even less clear than community, one I haven't tried to take up here, and one I'm not going to.) Despite the undeniable physical reality of it all, in the spring sun, I'm still not sure that I believe in the Kentlands exactly, but my visit makes it clear that someone does.

Notes

1 Introduction

1 Jacobs is a great favorite of the New Urbanists, in part for her opposition to Modernist urban renewal. It should be noted that Jacobs's work means many things to critics, planners, and designers from widely varying perspectives, thus cannot be understood to identify the New Urbanists ideologically or otherwise.

2 2004 was the last year for which *New Urban News* published a formal tabulation, one that they had been producing since 1995. The effort was apparently overwhelmed by definitional and information-gathering issues.

3 http://www.cnu.org/history, viewed June 18, 2009.

4 http://www.cnu.org/chapters, viewed June 18, 2009.

5 Elizabeth Plater-Zyberk has been Dean of the School of Architecture at the University of Miami since 1995; Douglas Kelbaugh has been Dean of the Taubman College of Architecture and Urban Planning at the University of Michigan since 1998; Stefanos Polyzoides recently declined the deanship of the University of Southern California School of Architecture; traditional designer Robert A.M. Stern, has been Dean of the Yale School of Architecture since 1998.

6 http://www.celebration.fl.us/towninfo.html; http://www.ramsa.com/project.aspx?id=27, both viewed July 24, 2010.

7 I owe the conception and execution distinction to Harry Braverman (1974).

8 In my understanding of the concept of field, I draw on the work of Pierre Bourdieu as elaborated below (Bourdieu 1985, 1990, 1996, 1998).

9 Note the connection to Latour's and Callon and Law's (1982) conceptions of interest: being impelled by the field is very much like the "consequences of whatever groups have been previously engaged to do," or "temporarily stabilized outcomes of previous processes of enrollment."

2 Before the New Urbanism

1 Films like *Poltergeist* (1982) could be construed as an implicit critique of suburban development, but it is not until the 1990s that the commentary becomes more explicit in films like *The Ice Storm* (1997), *Pleasantville* (1998), *The Truman Show* (1998), and *American Beauty* (1999).

2 This theory, along with many other aspects of the mortgage market, was of course put severely in doubt by the financial crisis of 2008, spurred primarily by the overvaluation of mortgage-backed securities and the underestimation of the risk they involved.

3 The legality of other, more directly related fees, stands.

4 Thanks to Susan Broeksmit for pointing this out (personal communication, September 9, 2006).

3 Making a movement

1 Polyzoides continued at Princeton for a Master's in architecture – where he met Kelbaugh – while Duany and Plater-Zyberk got their graduate degrees at Yale. Moule also got her Master's in architecture from Princeton, but is younger than the other three and did not overlap with them as students.

2 New Urban Timeline, nutimeline.net. Polyzoides was recently offered the deanship of the same school of architecture, but declined it to pursue his practice.

3 http://www.lgc.org/about/index.html#mission, viewed March 1, 2012.

4 Poticha did not make herself available for comment on her role in the CNU or the movement more generally, despite repeated requests.

5 I do not indentify this interviewee here or below because of their request in association with these particular comments.

6 As we will discuss in the next chapter, the New Urbanism's pragmatic engagement with the development market, as opposed to policy-led design, has been an essential part of its success.

7 http://www.nutimeline.net/, viewed January 16, 2006.

8 http://www.cnu.org/aboutcnu/index.cfm, viewed January 16, 2006.

9 http://www.cnu.org/charter, viewed March 1, 2012.

10 http://www.charretteinstitute.org/charrette.html, viewed January 17, 2006.

11 lakelands.org, viewed June 2, 2011.

12 Gaithersburg's Long Range Planning Director, Mark Depoe, indicated to me, in a tour of Kentlands and Lakelands, that one could tell the difference between the two developments by looking up at the roofing material. In Kentlands, cedar shingles are required, whereas in Lakelands, cheaper asphalt shingles are the norm.

4 Reimagining the suburbs, beginning to build

1 Various innovations in planning, like Planned Unit Developments, eased the transition between the two phases.

2 LEED is a widely used rating system for building design, construction, and operation, and LEED certification of a building guarantees a varying degree of energy and resource conservation. USGBC partnered with the CNU and the Natural Resources Defense Council (NRDC) to create a new standard LEED-ND or LEED for Neighborhood Development. Like the rating system for individual buildings, LEED-ND measures the conservation benefits associated with the location and spatial arrangement of a neighborhood-scale project. Andres Duany calls LEED-ND "the greatest tool." LEED-ND effectively consecrates many of the environmental claims of the New Urbanism by associating the CNU with USGBC, its rigorous reputation, and its popularity.

3 There are many more studies that explore the connections among the built environment, physical activity, and obesity, many of them indicting conventional suburban form.

4 Many retailers like drug store chains and supermarkets either obscure their windows with signage or use them for display, but obstruct views into and out of the body of the store itself [presentation at Providence CNU].

5 "Location Efficient Mortgage," locationefficiency.com, viewed June 20, 2007; Wikipedia, "Location Efficient Mortgage," en.wikipedia.org/wiki/Location_Efficient_Mortgage, viewed June 19, 2007.

6 Tu and Eppli (2001) show a range from a 4.1 per cent premium in the relatively unsuccessful Laguna West, near Sacramento, California, to a 14.9 per cent in Kentlands, by that year.

7 Ironically, the market success of many New Urbanist developments, by quickly driving prices out of reach of middle-income households, has brought charges of elitism from other critics, some of them previously or otherwise sympathetic. In response, New

Urbanists make a classic argument about supply and demand, suggesting that the more places they are allowed to build, the more prices will stabilize at a moderate level.

5 Surviving its critics and reinforcing its position

1 Among conservative publications, I reviewed and subcategorized the following: *Reason* magazine, the Reason Public Policy Institute's publications and web-pages, and *Policy Analysis* (libertarian); *The American Enterprise*, *Policy Review*, and *National Review* (mainstream); *The Weekly Standard*, *The Wall Street Journal*, and *Commentary* magazine (near-right).

2 Liberal publications reviewed included: *Mother Jones* magazine, *In These Times* (far-left); *The Nation*, *The New Yorker*, and the *American Prospect* (mainstream); *Slate Magazine* and *The New Republic* (TNR) (near-left). TNR is a long-time, major liberal press organ, but in recent years its ideological affiliation has become ambiguous.

3 And this note is not simply in their imagination, as we will see below: Duany is often identified as relatively conservative himself.

4 mississippi.org/content.aspx?url=/page/about&, viewed August 23, 2007.

5 mississippirenewal.com, viewed February 14, 2007.

6 By February 14, 2008, the FEMA "travel trailers" that were distributed around the Gulf Coast post-Katrina were discovered to off-gas significant quantities of formaldehyde into their interiors, causing levels of concentration that were both potentially cancerous and risked respiratory illnesses, according to a study by the CDC (FEMA 2008; www.cdc.gov/nceh/ehhe/trailerstudy/, viewed November 28, 2011).

Bibliography

[no author]. (2000) *The American Heritage dictionary of the English language*, Boston: Houghton Mifflin.

Abrams, Stanley D. (1991) "RE: Processing of Zoning Application Z–262(A)," Gaithersburg, Maryland.

Adler, J. (1995) "Bye-bye suburban dream," *Newsweek*, 15 May, pp. 40–45.

Andersen, Bengt. (2001) "The Quest for Community: A Study of Kentlands," Dissertation, Department of Social Anthropology, University of Oslo, Norway.

Arkin, Richard. (2004) "Kentlands in History," *Kentlands Telephone Book:* Winter 2004.

Bader, Michael. (2012) "Structuring Spatial Segregation through Place-Based Stigma," Presentation, Eastern Sociological Society Annual Meeting, Friday, February 24, 2012, New York, N.Y.

Barnes, Fred. (2000) "Suburban beauty: Why sprawl works," *The Weekly Standard*, May 22, 2000, pp. 27.

Bender, Thomas. (2003) "Scully's Way," *The Nation*, December 29, 2003, pp. 36–42.

Birnbaum, Charles. (2010) "City Shaping II: Will Architecture Go Horizontal?" *Huffington Post*. Online. Available HTTP: <http://www.huffingtonpost.com/mobileweb/1969/12/31/city-shaping-ii-will-arch_n_782812.html> (accessed November 1, 2011).

Blau, Judith R. (1984) *Architects and firms: a sociological perspective on architectural practice*, Cambridge, Mass.: MIT Press.

Blau, Judith, Mark La Gory, and John Pipkin. (1983) *Professionals and urban form*, Albany: State University of New York.

Bluestone, Barry and Bennett Harrison. (1982) *The deindustrialization of America: plant closings, community abandonment, and the dismantling of basic industry*, New York: Basic Books.

Boot, Max. (2002) "What the Heck Is a Neocon?" New York: *The Wall Street Journal*.

Bourdieu, Pierre. (1985) "The Social Space and the Genesis of Groups," *Theory and Society* 14: 723–44.

——. (1990) *The logic of practice*, Cambridge: Polity Press.

——. (1996) *The rules of art: genesis and structure of the literary field*, Cambridge: Polity Press.

——. (1998) *Practical reason: on the theory of action*, Stanford, Calif.: Stanford University Press.

Bradley, Jennifer. (2000) "Private Suburbs, Public Cities," *The American Prospect*, May 22, 2000, p. 46.

Brain, David. (2005a) "From good neighborhoods to sustainable cities: Social science and the social agenda of the new urbanism," *International Regional Science Review* 28: 217–38.

——. (2005b) "The politics of urbanism: pursuing a new urban ideal in a post-suburban world," Presentation at the American Sociological Association Annual Meeting, August 15, 2005, Philadelphia, PA.

Braverman, Harry. (1974) *Labor and monopoly capital: the degradation of work in the twentieth century*, New York: Monthly Review Press.

Brodkin, Karen. (1998) *How Jews became White folks: and what that says about race in America*, New Brunswick: Rutgers University Press.

Brown, Barbara B. and Vivian Cropper. (2001) "New urban and standard suburban subdivisions: Evaluating psychological and social goals," *Journal of the American Planning Association* 67: 402–20.

Bruegmann, Robert. (2005) *Sprawl: a compact history*, Chicago: University of Chicago Press.

——. (2006) "How Sprawl Got a Bad Name," *The American Enterprise*, June 2006, p. 16.

Callon, Michel and John Law. (1982) "On Interests and Their Transformation: Enrolment and Counter-Enrolment," *Social Studies of Law* 12: 615–25.

Calthorpe, Peter and Susan Benson. (1979) "The Solar Shadow: A Look at the Issues Eclipsed," Farallones Institute, Sacramento, California.

Cohen, Lizabeth. (2003) *A consumers' republic: the politics of mass consumption in postwar America*, New York: Knopf: Distributed by Random House.

Collier, Paul. (2009) "Beyond the begging bowl," *The Guardian*, April 2, 2009.

Conley, Dalton. (1999) *Being Black, living in the red*, Berkeley and Los Angeles: University of California Press.

Corbett, Judith, and Joe Velasquez. (1994) "The Ahwahnee Principles," *Western City Magazine*, 70, pp. 4–9.

Crane, R. and R. Crepeau. (1998) "Does neighborhood design influence travel? A behavioral analysis of travel diary and GIS data," *Transportation Research D* 3: 225–38.

Cuff, Dana. (1991) *Architecture: the story of practice*, Cambridge, Mass.: MIT Press.

Davis, Mike. (1990) *City of quartz: excavating the future in Los Angeles*, London; New York: Verso.

——. (2005) "Gentrifying Disaster," *Mother Jones*, October 25, 2005.

Davis, Mike and Anthony Fontenot. (2005) "25 Questions About the Murder of New Orleans," *The Nation*, September 30, 2005.

Dellinger, Matt. (2006) "The Terrible Opportunity: How a crisis for the Gulf Coast became a defining moment for New Urbanism," *Oxford American*, Spring 2006.

Duany, Andres, Elizabeth Plater-Zyberk, and Jeff Speck. (2000) *Suburban nation: the rise of sprawl and the decline of the American Dream*, New York: North Point Press.

Dunlop, Beth. (2001) "In Florida, a New Emphasis on Design," *The New York Times*. New York, N.Y.

Elliot, James R., Kevin Fox Gotham, and Melinda J. Milligan. (2004) "Framing the Urban: Struggles Over HOPE VI and New Urbanism in a Historic City," *City & Community* 3: 4.

Emrath, Paul. (2010) "Characteristics of Single-Family Homes Started in 2009," National Association of Home Builders.

Engebreth, Ben. (2005) "Housing Price-to-Income Ratio as a Way to Measure Maximum Local Affordability," Benengebreth.com, Thursday, June 2, 2005. Online. Available HTTP: <http://www.benengebreth.org/2005/06/housing_priceto> (accessed March 17, 2012).

——. (2012) "US Home Affordability," Department of Numbers, departmentofnumbers.com. Online. Available HTTP: <http://www.deptofnumbers.com/affordability/us/> (accessed March 17, 2012).

Eyal, Gil. (2002) "Dangerous Liasions between Military Intelligence and Middle Eastern Studies in Israel," *Theory and Society* 31: 653–693.

Fishman, Robert. (1987) *Bourgeois utopias: the rise and fall of suburbia*, New York: Basic Books.

——. (2000) *The American planning tradition: culture and policy*, Washington, D.C., Baltimore: Woodrow Wilson Centre Press; Johns Hopkins University Press.

Fornoff, Susan. (2006) "The little house that roared," *The San Francisco Chronicle*, March 4, 2006.

Frantz, Douglas, and Catherine Collins. (1999) *Celebration, U.S.A.: Living in Disney's Brave New Town*, New York: Henry Holt and Company.

Friedland, Roger. (1982) *Power and crisis in the city: corporations, unions, and urban policy*, London: Macmillan.

Gans, Herbert J. (1967) *The Levittowners; ways of life and politics in a new suburban community*, New York: Pantheon Books.

Giddens, Anthony. (1984) *The constitution of society: outline of the theory of structuration*, Cambridge: Polity Press.

Gladwell, Malcolm. (2000) *The tipping point: how little things can make a big difference*, Boston: Little, Brown.

Goldberger, Paul. (2000) "It takes a village: the anti-sprawl doctors make a manifesto," *The New Yorker*, March 27, 2000, pp. 128.

Goyette, Kimberly, John Iceland, and Elliot Weininger. (2012) "Moving for the Kids: Examining the Influence of Children on White Residential Segregation," Presentation, Eastern Sociological Society Annual Meeting, Friday, February 24, 2012, New York.

Greif, Mark. (2000) "Potemkin Villages," *The American Prospect*, January 3, 2000, p. 54.

Gross, Carson. (2000) "Comment," *Weekly Standard*, June 5, 2000.

Gudell, Svenja. (2011) "What Goes Up, Must Come Down: Comparing Price-to-Income Ratios Across Markets," Zillow Real Estate Research, Zillow.com, August 17, 2011. Online. Available HTTP: <http://www.zillow.com/blog/research/2011/08/17/what-goes-up-must-come-down-comparing-price-to-income-ratios-across-markets/ (accessed March 17, 2012).

Gutman, Robert. (1988) *Architectural practice: a critical view*, New York, N.Y.: Princeton Architectural Press.

Hales, Linda. (2006) "Katrina's Cottage Industry," *The Washington Post*, November 4, 2006, Washington, D.C.

Hall, Peter Geoffrey. (1988) *Cities of tomorrow: an intellectual history of urban planning and design in the twentieth century*, Oxford, U.K.; New York, N.Y., USA: Blackwell.

Hanchett, Thomas W. (1996) "U.S. Tax Policy and the Shopping-Center Boom of the 1950s and 1960s," *The American Historical Review*, Vol. 101, No. 4 (Oct., 1996), pp. 1082–110.

Handy, Susan. (2005) "Critical Assessment of the Literature on the Relationships Among Transportation, Land Use, and Physical Activity," Transportation Research Board and the Institute of Medicine Committee on Physical Activity, Health, Transportation, and Land Use.

Handy, Susan and Patricia Mokhtarian. (2005) "Which Comes First: The Neighborhood or the Walking?" *Access*, pp. 16–21.

Hanlon, James. (2010) "Success by design: HOPE VI, new urbanism, and the neoliberal transformation of public housing in the United States," *Environment and Planning A*, 42, pp. 80-98.

Harvey, David. (1997) "The New Urbanism and the Communitarian Trap," *Harvard Design Magazine*, Winter/Spring 1997. Cambridge, MA: Harvard Graduate School of Design.

Hawthorne, Christopher. (2005) "In the rush to rebuild, a house divided," *The Los Angeles Times*, Los Angeles, C.A.

Hayden, Dolores. (2003) *Building American suburbia: green fields and urban growth, 1820–2000*, New York: Pantheon Books.

Hayward, Steven. (1998) "Legends of the Sprawl," *Policy Review*, September/October 1998, pp. 26–31.

——. (1999) "Suburban Legends," *National Review*, March 22, 1999, pp. 35–8.

Henderson, Rick and Adrian T. Moore. (1998) "Plan Obsolescence," *Reason*, February 1998, pp. 42–47.

Hirsch, Arnold R. (1983/1998) *Making the Second Ghetto: race and housing in Chicago, 1940–1960*, Chicago: University of Chicago Press.

Hirt, Sonia A. (2009) "Premodern, Modern, Postmodern? Placing New Urbanism into a Historical Perspective," *Journal of Planning History* 8, pp. 248–73.

Hoch, Charles. (1994) *What planners do: power, politics, and persuasion*, Chicago, I.L.: Planners Press: American Planning Association.

Ikeda, Sandy. (2006) "Jane Jacobs, R.I.P." *National Review Online*, April 27, 2006.

Jackson, Kenneth T. (1985) *Crabgrass Frontier: The Suburbanization of the United States*, New York: Oxford University Press.

Jacobs, Jane. (1961) *The death and life of great American cities*, New York: Random House.

Journal of Blacks in Higher Education [The] (JBHE). (2011) "A Widening of the Black-White Income Gap," Breaking News, *The Journal of Blacks in Higher Education*, September 28, 2011. Online. Available HTTP: <http://www.jbhe.com/2011/09/a-widening-of-the-black-white-income-gap/> (accessed March 17, 2012).

Kamin, Blair. (2006) "Shortsighted Polemics," *Metropolis Magazine*. March 20, 2006.

Katz, Peter. (1994) *The New Urbanism: toward an architecture of community*, New York: McGraw-Hill, Inc.

Kay, Jane Holtz. (1998) *Asphalt nation: how the automobile took over America, and how we can take it back*, Berkeley: University of California Press.

Kelbaugh, Douglas, Ed. (1989) *The Pedestrian Pocket Book*, Princeton: Princeton Architectural Press.

Kim, Joongsub. (2001) "Sense of community in neotraditional and conventional suburban developments: A comparative case study of Kentlands and Orchard Village (Maryland)," PhD Dissertation Thesis, Architecture, Taubman College of Architecture and Urban Planning, University of Michigan, Ann Arbor, MI.

——. (2000) "Creating Community: Does the Kentlands Live up to Its Goals?" *Places*, 13: 2. Berkeley, C.A.

——. (2007) "Perceiving and Valuing Sense of Community in a New Urbanist Development: A Case Study of Kentlands," *Journal of Urban Design*, 12: 2, 203–30.

Kim, Joongsub, and Rachel Kaplan. (2004) "Physical and Psychological Factors in Sense of Community," *Environment and Behavior*, Vol. 36 No. 3, pp. 313–40.

Kunstler, James Howard. (1996) *Home from nowhere: remaking our everyday world for the twenty-first Century*, New York: Simon & Schuster.

Kruse, Kevin M., and Thomas J. Sugrue. (2006) *The New Suburban History*, Chicago: University of Chicago Press.

Langdon, Philip. (1994) *A better place to live: reshaping the American suburb*, Amherst: University of Massachusetts Press.

——. (1996) "The new, neighborly architecture," *The American Enterprise*, November/December 1996, pp. 41–46.

——. (2010) "After the Disaster," *New Urban Network*, March 1, 2010.

Larson, Magali Sarfatti. (1993) *Behind the postmodern facade: architectural change in late twentieth-century America*, Berkeley: University of California Press.
Latour, Bruno. (1987) *Science in action: how to follow scientists and engineers through society*, Cambridge, Mass.: Harvard University Press.
——. (1996) *Aramis, or, the love of technology*, Cambridge, Mass.: Harvard University Press.
Lazare, Daniel. (1997) "Mouseketopia," *In These Times*, March 17, 1997, p. 32.
Leccese, Michael, and Kathleen McCormick, Eds. (2000) *Charter of the New Urbanism*, New York: McGraw-Hill.
Lees, Loretta, Tom Slater, and Elvin Wyly. (2007) *Gentrification*: London and New York: Routledge.
Leigh, Catesby. (2003) "It Takes a (Well-Planned) Village: In praise of the New Urbanism," *National Review*, July 14, 2003.
——. (2006a) "'Live' with Witold Rybczynski," *The American Enterprise*, June 2006, pp. 11–15.
——. (2006b) "The Sins of Shady Lane," *The American Enterprise*, June 2006, pp. 22–26.
Leinberger, Christopher. (2000) "The Beginning of the End of Sprawl," *Urban Land*, January 2000, pp. 74–86.
——. (2004) "Financing New Urbanism," Congress for the New Urbanism XII, Chicago, Illinois.
——. (2005) "Retrofitting Real Estate Finance: Alternatives to the Nineteen Standard Real Estate Product Types," *Places*, Summer 2005.
Lewis, Jim. (2006) "Battle for Biloxi," *The New York Times*. New York, N.Y.
Liell, John Thomas. (1952) "Levittown: A Study in Community Planning and Development," Sociology, Yale University, New Haven, Connecticut.
Lindsay, Greg. (2010) "Is Haiti a Laboratory for New Urbanists? What the Country Really Needs Is Old Urbanism," *Fast Company*, April 6, 2010. Online. Available HTTP: <http://www.fastcompany.com/1607235/haiti-needs-old-urbanism-not-new> (viewed November 18, 2011).
Logan, John R. (1976) "Industrialization and the Stratification of Cities in Suburban Regions," *American Journal of Sociology* 82: 333–48.
Logan, John R. and Harvey Luskin Molotch. (1987) *Urban fortunes: the political economy of place*, Berkeley, C.A.: University of California Press.
Lopez, Kathryn Jean. (2006) "The Right Book," *National Review Online*, April 20, 2006.
Lubell, Sam. (2006) "Lowe's Makes Katrina Cottages Available for Purchase," *Architectural Record*, November 3, 2006.
Lund, H. (2003) "Testing the claims of new urbanism – Local access, pedestrian travel, and neighboring behaviors," *Journal of the American Planning Association* 69: 414–29.
McAdam, Doug. (1999) *Political process and the development of Black insurgency, 1930–1970*, Chicago: University of Chicago Press.
MacCash, Doug. (2009) "Urban Planner Andres Duany shows off his Bywater house prototypes," *The Times Picayune*, January 31, 2009. Viewed at Nola.com, November 18, 2011.
McKenzie, Evan. (1994) *Privatopia: homeowner associations and the rise of residential private government*, New Haven: Yale University Press.
Mareth, Joanna. (2001) "Faux Urbanism," *The American Prospect*, June 18, 2001, pp. 45.
Marshall, Alex. (1996) "Putting Some 'City' Back In the Suburbs; Sounds Interesting – Too Bad It Doesn't Work," *The Washington Post*, pC01.
Martin, John Levy. (2003) "What Is Field Theory?" *The American Journal of Sociology* 109: 1–49.

Massing, Michael. (2000) "Stalled in Paradise," *The American Prospect*, May 22, 2000, p. 22.

Mayor and City Council. (1989) "Minutes of a Regular City Council Meeting," Gaithersburg, Maryland.

Mehan, G. Tracy, III. (2006) "Keeping Cities Free," *National Review Online*, April 27, 2006.

Mississippi Renewal Forum. (2005) "A Pattern Book for Gulf Coast Neighborhoods," Pittsburgh: Urban Design Associates.

Mohney, David, and Keller Easterling, Eds. (1991) *Seaside: making a town in America*, New York: Princeton Architectural Press.

Molotch, Harvey Luskin. (2003) *Where stuff comes from: how toasters, toilets, cars, computers, and many others things come to be as they are*, New York: Routledge.

Morrow-Jones, Hazel A., Elena G. Irwin, and Brian Roe. (2004) "Consumer Preference for Neotraditional Neighborhood Characteristics," *Housing Policy Debate* 15: 171–202.

Muschamp, Herbert. (1994) "Architecture View: A Show Brings On a Severe Case Of Period Envy," *The New York Times*, p40.

——. (1996) "Can New Urbanism Find Room for the Old?" *The New York Times*. June 2, 1996, p 40.

——. (2000) "Throwing Our Anxieties to the Lions," *The New York Times*, April 30, 2000.

Nasar, Jack L. (2003) "Does Neotraditional Development Build Community?" *Journal of Planning Education and Research* 23, pp. 58–68.

New Urban News. (2003) "New Urban Projects on a Neighborhood Scale in the United States," *New Urban News*.

O'Toole, Randal. (1999) "Dense Thinkers," *Reason*, January 1999, pp. 44–52.

Pahl, R. E. (1969) "Urban social theory and research," *Environment and Planning* 1(2) pp. 143–153.

Patterson, Robert W. (2000) "Comment," *Weekly Standard*, June 5, 2000.

Pickering, Andrew. (1995) *The mangle of practice: time, agency, and science*, Chicago: University of Chicago Press.

Planning Commission, J. Russel, ed. (1991a) "Memo to Mayor and City Council re: Z–262(A)," Gaithersburg, Maryland.

Planning Commission. (1991b) "Planning Commission Minutes: Z–262(A)," Gaithersburg, Maryland.

Planning Commission and Mayor and City Council. (1988) "Map Amendment Application Z–262," Gaithersburg, Maryland.

Planning Commission and Mayor and City Council. (1991) "Map Amendment Z–262(A)," Gaithersburg, Maryland.

Plaut, Pnina O., and Marlon G. Boarnet. (2003) "New Urbanism and the Value of Neighborhood Design," *Journal of Architectural and Planning Research* 20: 254–65.

Pogrebin, Robin. (2006) "An architect with plans for a new Gulf Coast," *The New York Times*, May 24, 2006.

Popkin, Susan J., Bruce Katz, Mary K. Cunningham, Karen D. Brown, Jeremy Gustafson, and Margery A. Turner. (2004) "A DECADE OF HOPE VI: Research Findings and Policy Challenges," The Urban Institute. Washington, D.C.

Putnam, Robert D. (2000) *Bowling alone: the collapse and revival of American community*, New York: Simon & Schuster.

Pyatok, Michael. (2000) "Comment on Charles C. Bohl's 'New Urbanism and the City: Potential Applications and Implications for Distressed Inner-City Neighborhoods'—The Politics of Design: The New Urbanists vs. the Grass Roots," *Housing Policy Debate*, 11: 4.

Real Estate Research Corporation. (1974) *The Costs of Sprawl*, Washington, D.C.: U.S. Government Printing Office.

Riesman, David. (1961) *The lonely crowd; a study of the changing American character*, New Haven: Yale University Press.

Risen, Clay. (2005) "Wrong Way Home," *The Morning News*, November 8, 2005. Online. Available HTTP: <http:// www.themorningnews.org/article/wrong-way-home> (viewed November 10, 2011).

Rood, T. (2003) "Community by design: New urbanism for suburbs and small communities," *Journal of the American Planning Association* 69: 323–4.

Ross, Andrew. (1999) *The Celebration chronicles: life, liberty, and the pursuit of property value in Disney's new town,* New York: Ballantine Books.

Rybczynski, Witold. (1995) "The Rise of Family Values Architecture," *The New Republic*, May 8, 1995, p. 14.

——. (2006) "How To Rebuild New Orleans: Start with a visit to Denver," *Slate*, August 24, 2006. Online. Available HTTP: <http://www.slate.com/articles/arts/architecture/2006/08/how_to_rebuild_new_orleans.html (accessed November 1, 2006).

Scott, James C. (1998) *Seeing Like a State: how certain schemes to improve the human condition have failed*, New Haven: Yale University Press.

Seelye, Katharine Q. (2007) "New Republic Cuts Back, but Bulks Up Its Image," *The New York Times*, New York, N.Y.

Shoskes, Ellen. (1989) *The design process: an illustrated guide for architects and interior designers*, New York: Whitney Library of Design.

Skocpol, Theda. (2003) *Diminished democracy: from membership to management in American civic life*, Norman: University of Oklahoma Press.

Skocpol, Theda and Morris P. Fiorina. (1999) *Civic engagement in American democracy*, Washington, D.C., New York: Brookings Institution Press; Russell Sage Foundation.

Song, Yan and Gerrit-Jan Knaap. (2003) "New urbanism and housing values: a disaggregate assessment," *Journal of Urban Economics* 54: 218–38.

——. (2004) "Measuring Urban Form: Is Portland Winning the War on Sprawl?" *Journal of the American Planning Association* 70: 210–25.

Sorkin, Michael. (2006) "Will new plans for the Gulf drown it again, this time in nostalgia?" *Architectural Record*, February 2006, pp. 47.

Southworth, M. (1997) "Walkable suburbs? An evaluation of neotraditional communities at the urban edge," *Journal of the American Planning Association* 63: 28–44.

Staley, Sam. (2001) "Room to Grow," *Reason*, February 2001, pp. 48–54.

Stark, Judy. (2006) "The house that Katrina built," *The St. Petersburg Times*, January 28, 2006.

Steigerwald, Bill. (2001) "City Views," *Reason*, June 2001, pp. 48–55.

Stein, Jeff. (2010) "Groundswell: The rise of landscape urbanism," ArchitectureBoston, Fall 2010, Boston, MA. Online. Available HTTP: <http://architectureboston.wordpress.com/2010/08/04/groundswell/> (accessed November 15, 2012).

Steuteville, Robert. (2005) "New Urbanism continued strong growth in 2004," *New Urban News*, January/February 2005.

Steuteville, Robert, and Philip Langdon. (2003) *New Urbanism: Comprehensive Report and Best Practices Guide*, Ithaca, N.Y.: New Urban Publications.

Talen, Emily. (1999) "Sense of community and neighbourhood form: An assessment of the social doctrine of new urbanism," *Urban Studies* 36:1361–79.

——. (2005) *New Urbanism and American Planning*, New York: Routledge.

——. (2010) "Affordability in New Urbanist Development: Principle, Practice and Strategy." *Journal of Urban Affairs*, 32: 4, pp.489–510.

Taylor, Paul, Rakesh Kochhar, Richard Fry, Gabriel Velasco, and Seth Motel. (2011) "Twenty-to-One: Wealth Gaps Rise to Record Highs Between Whites, Blacks and Hispanics," Pew Research Center. Washington, D.C.

The Week. (2011) "America's traffic gridlock: By the numbers," January 21, 2011. Online. Available HTTP: <http://theweek.com/article/index/211347/americas-traffic-gridlock-by-the-numbers> (accessed November 15, 2011).

Tu, Charles C. and Mark J. Eppli. (1999) "Valuing New Urbanism: The Case of Kentlands," *Real Estate Economics* 27: 425–451.

——. (2001) "An Empirical Examination of Traditional Neighborhood Development," *Real Estate Economics* 29: 485–501.

U.S. Department of Homeland Security, Federal Emergency Management Agency (FEMA). (2008) "CDC Releases Results Of Formaldehyde Level Tests: FEMA To Expedite Relocation Of Residents From Temporary Housing Units," Release Date: February 14, 2008, Release Number: HQ-08-021.

U.S. Department of Housing and Urban Development (HUD). (ND) "About HOPE VI." Online. Available HTTP: <http://portal.hud.gov/hudportal/HUD?src=/program_offices/public_indian_housing/programs/ph/hope6/about> (viewed November 14, 2011).

U.S. Department of Housing and Urban Development and Congress for the New Urbanism. (2000) "Principles for Inner City Neighborhood Design: HOPE VI and the New Urbanism," Washington, DC: U.S. Government Printing Office.

Waldheim, Charles. (2010) Presentation at the University of North Carolina College of Arts and Architecture, Charlotte, NC, February 17, 2010. http://nexturbanism.posterous.com/charles-waldheim-unc-lecture–2010 (accessed October 24, 2011).

Weiss, Marc A. (1987) *The rise of the community builders: the American real estate industry and urban land planning*, New York: Columbia University Press.

Whyte, William Hollingsworth. (1956) *The organization man*, New York: Simon & Schuster.

Zinsmeister, Karl. (1996) "Are today's suburbs really family-friendly?" *The American Enterprise*, November/December 1996, pp. 36–41.

——. (2006) "Bird's Eye: In praise of ordinary choices," *The American Enterprise*, June 2006, p. 4.

Index